GUIDE TO VEGETABLE GARDENING AND PRESERVING

McGraw-Hill Paperbacks
Home Improvement Series

Guide to Plumbing
Guide to Electrical Installation and Repair
Guide to Roof and Gutter Installation and Repair
Guide to Wallpaper and Paint
Guide to Paneling and Wallboard
Guide to Landscape and Lawn Care
Guide to Brick, Concrete, and Stonework
Guide to Carpentry
Guide to Furniture Refinishing and Antiquing
Guide to Bathroom and Kitchen Remodeling
Guide to Reupholstering
Guide to Vegetable Gardening and Preserving
Guide to Bicycle Repair and Maintenance
Guide to Floor and Carpeting Installation and Repair

GUIDE TO VEGETABLE GARDENING AND PRESERVING

McGraw-Hill Book Company

New York St. Louis San Francisco Auckland Bogotá Guatemala Hamburg
Johannesburg Lisbon London Madrid Mexico Montreal New Delhi
Panama Paris San Juan São Paulo Singapore Sydney Tokyo Toronto

1 2 3 4 5 6 7 8 9 0 SMSM 8 6 5 4 3 2 1

Library of Congress Cataloging in Publication Data

Main entry under title:

Guide to vegetable gardening and preserving.

(McGraw-Hill paperbacks home improvement series)
Previous ed. (c1976) by Minnesota Mining and Manufacturing Company, Hardware-Paint Trades Division, published under title: The home pro vegetable gardening and preserving guide.
1. Vegetable gardening. 2. Vegetables — Preservation. I. Minnesota Mining and Manufacturing Company. Hardware-Paint Trades Division. The home pro vegetable gardening and preserving guide.
SB321.G86 635 80-18102
ISBN 0-07-045969-X (pbk.)

Cover photo by Judith Mallinson

Contents

HOW TO USE THIS BOOK

For any of the jobs in this book, you'll find detailed step-by-step instructions. The steps are completely illustrated to show exactly what is meant by the instruction.

Before beginning any job, it is a good idea to first read through the entire procedure. You will be told the equipment, tools and supplies you need. Also, you will get a pretty good idea of just what skills and time are involved.

Once you begin the job, proceed step-by-step. Be sure to read any information in front of the step. Because the step is keyed directly to the illustration, learn to refer to the illustration while you are reading. It may be slower reading but it's faster understanding.

GUIDE TO VEGETABLE GARDENING AND PRESERVING

OUTDOOR VEGETABLE GARDENS

PLANNING THE GARDEN

Planning a vegetable garden is different than planning a permanent structure, such as adding a room. The vegetable garden can be changed relatively easily, as the home gardener gets more experience. It is normal for home gardeners to experiment with different types of vegetables. It is also not unusual to make the garden larger or smaller, or even change its location.

Because of this flexibility, it is not necessary to decide on a fixed and final plan on the first attempt. And this is very fortunate. Successful home vegetable gardening depends on general rules properly mixed with specific knowledge and experience of the individual gardener. Climate and soil can change markedly within fairly small geographical regions, and will have considerable effect on your selection of vegetables to plant. The home gardener should therefore view the planning suggestions presented here mainly as guidelines, and proceed in an experimental frame of mind.

In planning a vegetable garden, you will first want to select the types of vegetables to plant. You will be better able to make this selection if you first become familiar with the different types of vegetables from the standpoint of growing, rather than eating. Of course, what you like to eat should and will influence your choice of vegetables to grow.

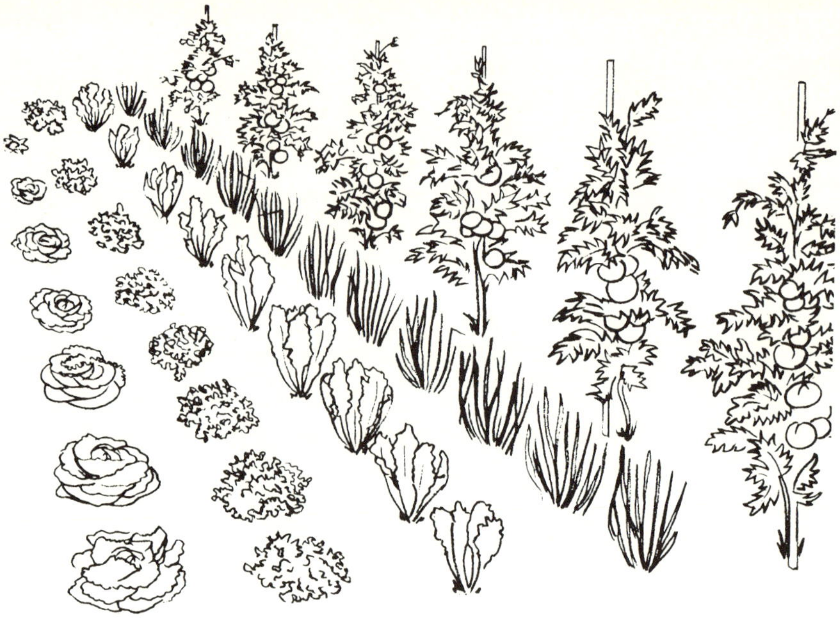

Size and location of the garden depend on a number of factors. These factors will also be discussed before suggesting ways of laying out your garden.

▶ Types of Vegetables

Vegetables can be grouped into two types: warm season vegetables and cool season vegetables. This is an important distinction for the home gardener, because it can affect both what to plant and when to plant.

Of the major factors affecting vegetable growth, such as climate, soil, and water, climate is the least under control. If the soil is bad, it can be improved. If the area has little rainfall, irrigation water can be supplied. But, lacking a greenhouse, the home gardener is at the mercy of the local climate. By careful attention to the time of planting, however, even the climatic problems can be solved to a large extent.

The warm season vegetables are mainly those which bear a "fruit": beans, corn, cucumbers, eggplant, melons, peppers, squash, and tomatoes.

WARM SEASON VEGETABLES

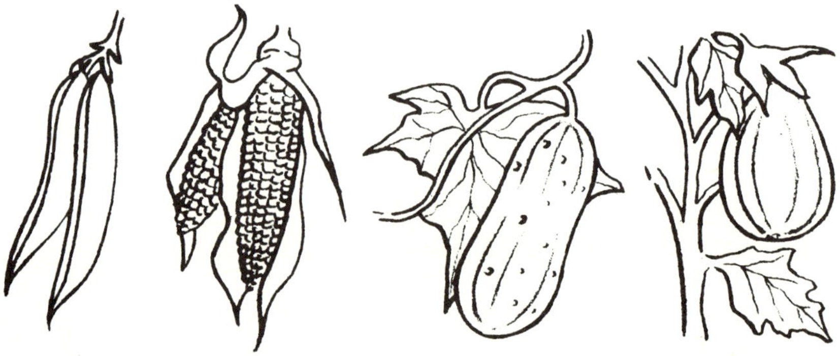

BEANS CORN CUCUMBERS EGGPLANT

MELONS PEPPERS SQUASH TOMATOES

COOL SEASON VEGETABLES

Types of Vegetables

The cool season vegetables tend to be leafy or root vegetables: asparagus, beets, broccoli, brussels sprouts, cabbage, carrots, cauliflower, celery, chard, lettuce, onions, peas, potatoes, radishes, spinach, and turnips.

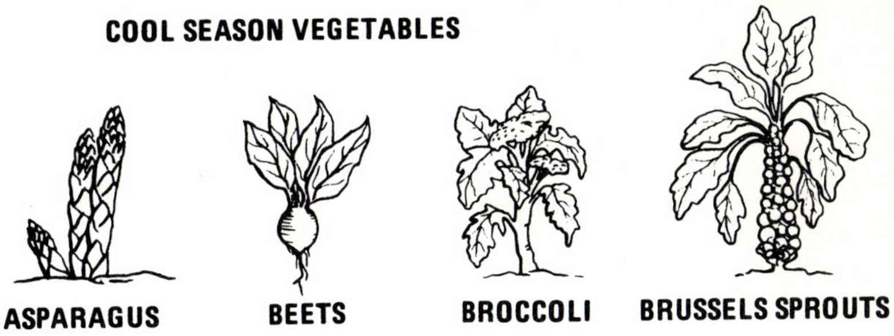

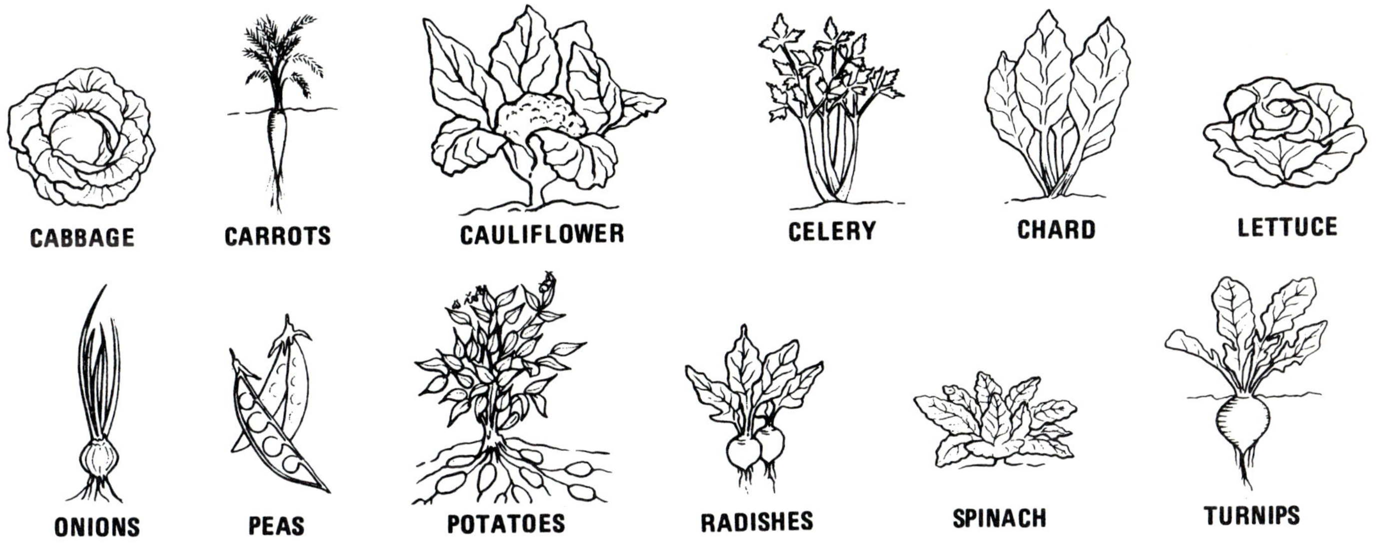

Types of Vegetables

There are no rigid rules which match specific vegetables to specific locations or climates. Most of the vegetables discussed in this guide can be successfully grown in most parts of the United States, if planted at the right time. If you live in an area with continually cool ($55^{o} - 70^{o}$F) climate, however, you may have trouble growing warm season vegetables. Or if you have a very short cool season between severe winters and hot summers, the growing time for certain cool season vegetables may be marginal.

The situation is as follows: warm season vegetables require a period of warm ($65^{o} - 85^{o}$F) weather to grow properly. Some vegetables require a longer time than others, as described later. If the warm season vegetables do not experience a long enough period of warm weather, any of several things may occur: (1) the seeds may not sprout, (2) the plants may not grow, or (3) the "fruit" may not form.

Cool season vegetables require a period of cool ($55^{o} - 70^{o}$F) weather to grow properly. If the weather becomes warm too soon, the cool season vegetables will produce premature flowers and seeds.

Whether warm season type or cool season type, most vegetables are sensitive to frost, some more so than others. The growing period is considered to extend from the last spring frost to the first fall frost. It is during this period that most vegetables are planted. Also, the temperatures during this growing period will determine the behavior of warm season and cool season vegetables.

The table on the next page gives the average dates of the last spring frost and first fall frost in different geographical areas. Find the location nearest you, in order to see what your average growing period is and when frosts occur. The dates are averages only. You should allow about ten days either side for local and seasonal variations.

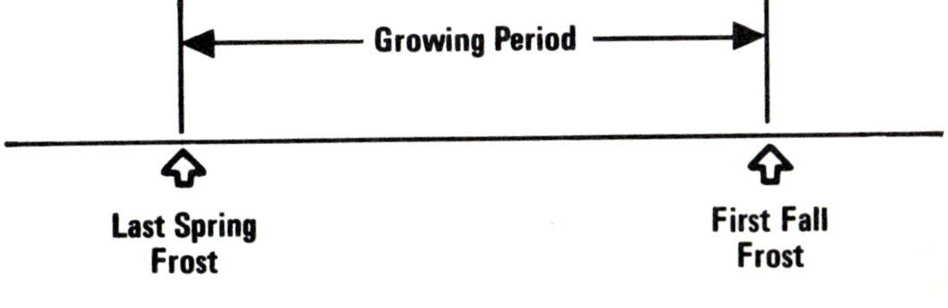

FROST DATES BY GEOGRAPHICAL AREA

State	Last in Spring	First in Fall
Alabama		
Northwest	Mar 25	Oct 30
Southeast	Mar 8	Nov 15
Arizona		
North	Apr 23	Oct 19
South	Mar 1	Dec 1
Arkansas		
North	Apr 7	Oct 23
South	Mar 25	Nov 3
California		
Imperial Valley	Jan 25	Dec 15
Interior Valley	Mar 1	Nov 15
Southern Coast	Jan 15	Dec 15
Central Coast	Feb 25	Dec 1
Mountain Sections	Apr 25	Sept 1
Colorado		
West	May 25	Sept 18
Northeast	May 11	Sept 27
Southeast	May 1	Oct 15
Connecticut	Apr 25	Oct 20
Delaware	Apr 15	Oct 25
District of Columbia	Apr 11	Oct 23
Florida		
North	Feb 25	Dec 5
Central	Feb 11	Dec 28
South*		
Georgia		
North	Apr 1	Nov 1
South	Mar 15	Nov 15
Idaho	May 21	Sept 22

State	Last in Spring	First in Fall
Illinois		
North	May 1	Oct 8
South	Apr 15	Oct 20
Indiana		
North	May 1	Oct 8
South	Apr 15	Oct 20
Iowa		
North	May 1	Oct 2
South	Apr 15	Oct 9
Kansas	Apr 20	Oct 15
Kentucky	Apr 15	Oct 20
Louisiana		
North	Mar 13	Nov 10
South	Feb 20	Nov 20
Maine	May 25	Sept 25
Maryland	Apr 19	Oct 20
Massachusetts	Apr 25	Oct 25
Michigan		
Upper Peninsula	May 25	Sept 15
North	May 17	Sept 25
South	May 10	Oct 8
Minnesota		
North	May 25	Sept 15
South	May 11	Oct 1
Mississippi		
North	Mar 25	Oct 30
South	Mar 15	Nov 15
Missouri	Apr 20	Oct 20
Montana	May 21	Sept 22

State	Last in Spring	First in Fall
Nebraska		
West	May 11	Oct 4
East	Apr 15	Oct 15
Nevada		
West	May 19	Sept 22
East	June 1	Sept 14
New Hampshire	May 23	Sept 25
New Jersey	Apr 20	Oct 25
New Mexico		
North	Apr 23	Oct 17
South	Apr 1	Nov 1
New York		
West	May 10	Oct 8
East	May 1	Oct 15
North	May 15	Oct 1
No. Carolina		
West	Apr 15	Oct 25
East	Apr 8	Nov 1
No. Dakota		
West	May 21	Sept 13
East	May 16	Sept 20
Ohio		
North	May 6	Oct 15
South	Apr 20	Oct 20
Oklahoma	Apr 2	Nov 2
Oregon		
West	Apr 17	Oct 25
East	June 4	Sept 22

State	Last in Spring	First in Fall
Pennsylvania		
West	Apr 20	Oct 10
Central	May 1	Oct 15
East	Apr 17	Oct 15
Rhode Island	Apr 25	Oct 25
So. Carolina		
Northwest	Apr 1	Nov 8
Southeast	Mar 15	Nov 15
So. Dakota	May 15	Sept 25
Tennessee	Apr 10	Oct 25
Texas		
Northwest	Apr 15	Nov 1
Northeast	Mar 21	Nov 10
South	Feb 10	Dec 15
Utah	Apr 26	Oct 19
Vermont	May 23	Sept 25
Virginia		
North	Apr 15	Oct 25
South	Apr 10	Oct 30
Washington		
West	Apr 10	Nov 15
East	May 15	Oct 1
W. Virginia		
West	May 1	Oct 15
East	May 15	Oct 1
Wisconsin		
North	May 17	Sept 25
South	May 1	Oct 10
Wyoming		
West	June 20	Aug 20
East	May 21	Sept 20

*Almost frost free

Types of Vegetables

Early and late varieties are available for both warm season and cool season vegetables. These varieties were developed to plant at different times. However, they also give you a wider choice of vegetables for your climate. For example, if your warm period is short, early varieties of warm season vegetables, which require less heat, may grow well. Or if your cool period is short, a late variety of cool season vegetable may be best.

Different varieties are described in seed catalogs, which are available by mail order. A few of the larger vegetable seed companies and their addresses are listed at the right.

If you plan to preserve your vegetables by freezing, canning, etc., you may wish to select specific varieties which are best for this purpose. Some of these varieties are listed in the section on Preserving Vegetables (Page 57).

MAJOR SEED COMPANIES

1. Burgess Seed & Plant Co.
P.O. Box 218
Galesburg, MI 49053
2. W. Atlee Burpee
Philadelphia, PA 19132
Clinton, IA 52732
Riverside, CA 92502
3. Farmer Seed & Nursery
Faribault, MN 55021
4. Henry Field Seed & Nursery Co.
Shenandoah, IA 51601
5. Gurney Seed & Nursery Co.
Yankton, SD 57078
6. Joseph Harris Co., Inc.
Moreton Farm
Rochester, NY 14642
7. J.W. Jung Co.
Randolph, WI 53956
8. Earl May Seed & Nursery Co.
Shenandoah, IA 51601
9. R.H. Shumway
P.O. Box 777
Rockford, IL 61101

▶ Characteristics of Different Vegetables

In planning a vegetable garden, it is helpful to know more about the individual vegetables—their characteristics and requirements. Some vegetables require more care than others, some take up more space than others, some cannot be grown in a hot climate, etc. Most of the details of planting and caring for vegetables will be covered later, but it is useful to understand the more basic characteristics during this planning stage.

One of the most important characteristics of individual vegetables is, of course, taste. Words and pictures will be of no help here; each gardener must rely on his or her own experiences. However, a word of caution is needed. Many home gardeners agree that vegetables taste better—and in some cases different—when home-grown. So before excluding too many vegetables based on past taste experiences, consider giving them another chance. You may be pleasantly surprised!

A number of the basic characteristics useful for planning are summarized in the table on the following pages. Down the left side are listed 24 vegetables commonly grown in the United States. There are others that you may wish to grow depending on your tastes and location.

Across the top of the table are some of the basic characteristics. In order, from left to right, these are as follows:

Cool or warm season: This is a basic characteristic discussed previously. It is a guide both as to when to plant and possibly, depending on climate, whether you should plant at all. Cool season vegetables generally like a daytime growing temperature of $55^{\circ}-70^{\circ}$F, and are tolerant of some frost. These temperatures are approximate only; cool season vegetables may be grown in conditions outside this range, but usually with less success.

Warm season vegetables generally like day and night temperatures of $65^{\circ}-80^{\circ}$F. Again, these are approximate only. Warm season vegetables are easily damaged by frost. Because of the different temperature requirements, cool season vegetables are usually planted in early spring or in late summer or in fall. Warm season vegetables are usually planted in late spring or in summer.

Planting form: Vegetables are normally planted either from seeds or in the form of seedlings. Many are planted both ways. The most common form is shown in the table. Where "seedlings" is the entry in the table, the home gardener may wish to grow seedlings indoors from seed. How this is done is described later (Page 31).

Time to harvest: This is the time to mature from the planting form (e.g., seeds) shown. This shows how long you must wait before eating the vegetable. It can also help in planning. Success or failure of an individual vegetable in your climate will depend on the temperature requirements (cool or warm season) over the growing period (time to harvest). For example, if you have a short summer, any warm season vegetables you choose should have a short time to harvest, to correspond to your short summer.

Succession planting: Many vegetables, particularly faster growing ones, are often planted more than once during a season. For example, carrots are typically planted at two-week intervals, so as to provide a continuous harvest later. In planning the garden, space must be provided (as discussed later, Page 20) for succession planting if this is the intention.

Fertilizer requirements: Some vegetables can be planted in well-fertilized soil (as described in Preparing the Soil, Page 21) and need no additional fertilizer throughout their growth. Other vegetables will do best if there is fertilizer added periodically. These latter vegetables, which require this additional effort, have an entry of "heavy" in the table.

Water requirements: A vegetable garden is usually watered uniformly, as a unit (see Watering, Page 35). While all vegetables require water during their growth, some require more than others. These will be the first affected if the garden as a whole is not getting enough water. This is indicated in the table by the word "frequent" opposite these vegetables, to show the extra attention that may be required.

Thinning requirements: When vegetables are planted in the form of seeds, later "thinning" is usually beneficial. Some of the small plants are removed to provide more room for the others. If seedlings are grown at home instead of bought, thinning will also be required (but this is not shown in the table).

Space requirements: Knowing the approximate size of vegetable plants is useful in allocating growing space in the garden, or even in deciding whether or not to grow some particular item. The space given to some of the larger vegetables, such as cucumbers, can be reduced by training the vegetables vertically up supports. Small, fast-growing vegetables can be planted among larger plants, so that no separate growing space is required (see Intercropping, Page 20).

Support requirements: Some vegetable plants, particularly the larger ones, are usually supported by some sort of structure such as wooden stakes, fencing, etc. This reduces the area needed for growing and keeps the vegetables off the ground.

Remarks: This is a "reminder" column which includes some of the essential characteristics discussed later. It is placed here to help round out the picture.

PLANNING THE GARDEN

Characteristics of Different Vegetables

VEGETABLE CHARACTERISTICS

Vegetable	Cool or Warm Season	Planting Form	Time to Harvest	Succession Planting	Fertilizer Req.	Water Req.	Thinning Req.	Space Req.	Support Req.	Remarks
Asparagus	Cool	Crowns	2 yrs.	No	Twice/yr.	Normal	No	Med. to large	No	Perennial
Beans	Warm	Seeds	Bush - 50 days Pole - 65 days	Yes	Normal to heavy	Normal to frequent	Yes	Bush - med Pole - large	Bush - no Pole - yes	Easily killed by frost
Beets	Cool	Seeds	60 days	Yes	Normal	Normal to frequent	Yes	Small	No	Tolerates wide temperature range
Broccoli	Cool	Seedlings	60–80 days	Yes	Heavy	Normal	No	Med.	No	Easy to grow, sensitive to heat
Brussels Sprouts	Cool	Seedlings	90 days	No	Heavy	Normal	No	Med.	No	Very productive
Cabbage	Cool	Seedlings	60–90 days	Yes	Heavy	Normal to frequent	No	Med.	No	Rotate crops
Carrots	Cool	Seeds	60 days	Yes	Normal	Normal	Yes	Small	No	Keep seeds moist to germinate
Cauliflower	Cool	Seedlings	60–80 days	No	Normal	Normal	No	Med.	No	Sensitive to frost, blanch head
Celery	Cool	Seedlings	120 days	Yes	Heavy	Frequent	No	Small	No	Difficult to grow
Chard	Cool	Seeds	60 days	No	Normal	Normal	Yes	Small	No	Can cut continually
Corn	Warm	Seeds	75–90 days	Yes	Heavy	Normal to frequent	Yes	Med. to large	No	Large number of varieties; cook immediately
Cucumber	Warm	Seeds	50–70 days	No	Normal	Frequent	Yes	Med. to large	Yes	Sensitive to frost
Eggplant	Warm	Seedlings	80 days	No	Heavy	Frequent	No	Med.	No	Rotate crops Decorative
Lettuce	Cool	Seeds	45–80 days	Yes	Normal	Frequent	Yes	Small	No	Leaf lettuce grows continuously
Melons	Warm	Seeds or seedlings	60–120 days	No	Heavy	Frequent	Yes (seeds)	Large	Depends on space	Many varieties
Onions	Cool	Seeds, sets, plants	100–120 days	No	Normal	Frequent	Yes (seeds)	Small	No	Hardy
Peas	Cool	Seeds	60–70 days	Yes	Normal	Frequent	No	Med. to large	Yes	Cook immediately Bushes or vines
Peppers	Warm	Seedlings	70–80 days	No	Heavy	Frequent	No	Med.	No	Decorative
Potatoes	Cool	Sets	75–120 days	No	Normal	Normal	No	Med. to large	No	Obtain sets by mail-order
Radishes	Cool	Seeds	20–30 days	Yes	Normal	Normal	Yes	Small	No	Easy to grow
Spinach	Cool	Seeds	45 days	Yes	Normal to heavy	Normal	Yes	Small	No	Cook or eat raw
Squash	Warm	Seeds or seedlings	60–120 days	No	Heavy	Normal	Yes (seeds)	Large	Depends on space	Summer squash: easy to grow Winter squash: long growing season
Tomatoes	Warm	Seedlings	60–80 days	No	Normal	Normal	No	Med.	Yes	Rotate crops
Turnips	Cool	Seeds	45–60 days	Yes	Normal	Frequent	Yes	Small	No	Eat tops and roots

Characteristics of Different Vegetables

The individual vegetables can now be discussed briefly in terms of these basic characteristics. In addition, any other information useful for planning will be mentioned.

1. Asparagus

This is a perennial plant: plant it once and it keeps on producing periodically for 10–15 years. It is the only perennial on the list of 24 vegetables. Because of its permanence, it needs an area off to the side, by itself. Its attractive appearance makes it useful for decoration.

Asparagus is planted in the form of "crowns" (roots). Even with crowns, there is no production for two years. Good, steady production begins in the third year.

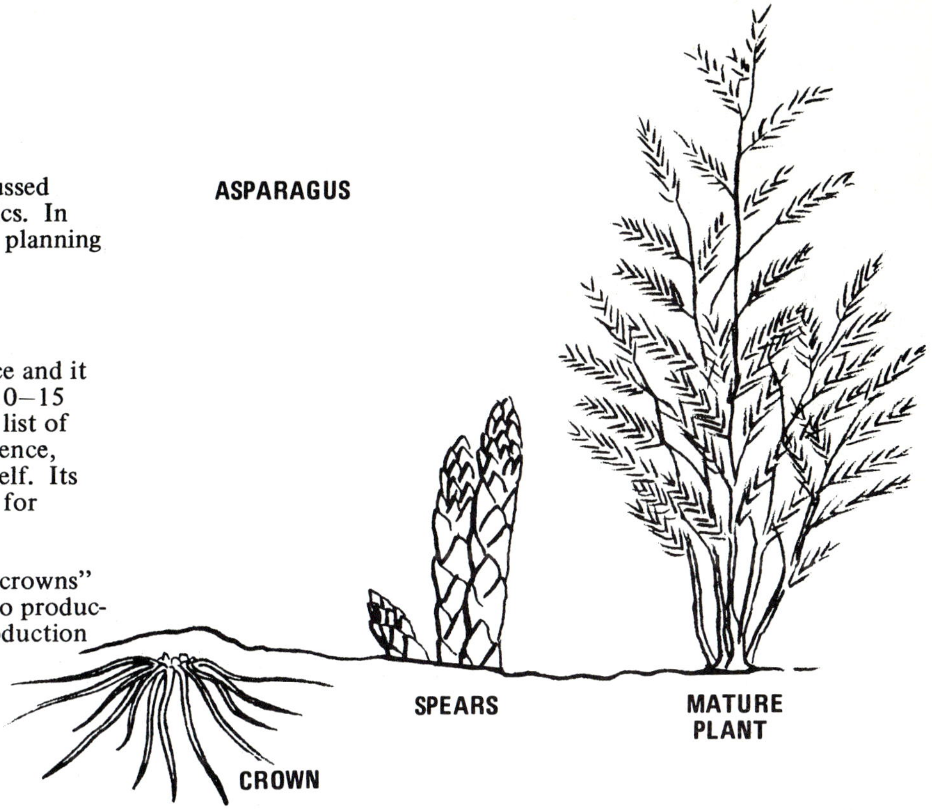

Characteristics of Different Vegetables

2. Beans

There are two main types of beans—snap (also called green, string, or wax beans) and lima (also called butterbeans). Each of these, snap and lima, comes in two forms, bush and pole. Bush beans are compact and require no support. Pole beans are considerably larger and are trained to grow vertically on supports. Bush beans mature faster than pole beans—about 50 days compared to 65 days.

Beans help improve the fertility of the soil. They require substantial amounts of fertilizer and water.

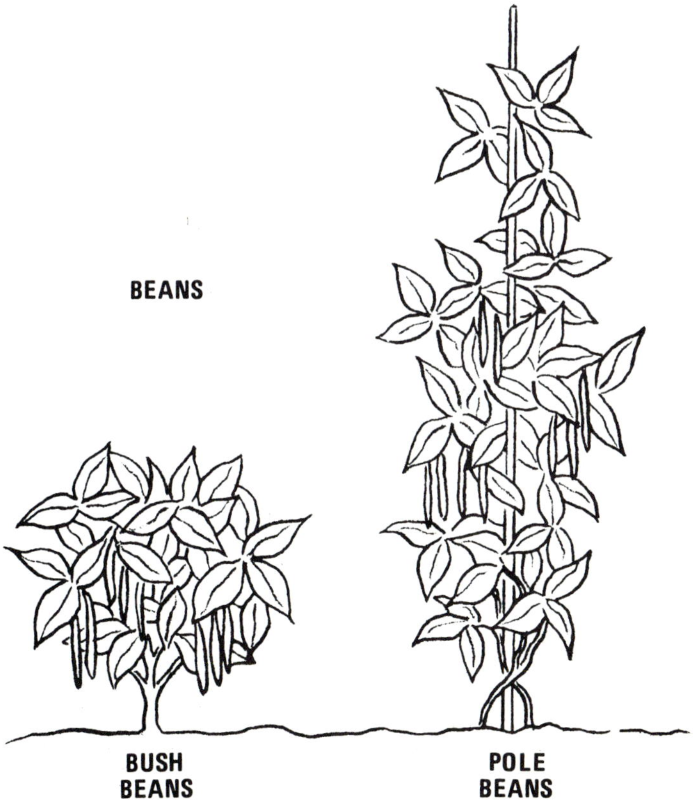

PLANNING THE GARDEN

Characteristics of Different Vegetables

3. Beets

Beets are easy to grow and provide a relatively large output for a given area. They grow rapidly and are tolerant of a fairly wide temperature range. Both the roots and the tops can be eaten.

4. Broccoli

Broccoli is easy to grow but is very sensitive to heat. If the weather is too warm it will produce flowers. Broccoli is a member of the cabbage family, all of which like ample fertilizer.

5. Brussels Sprouts

Brussels sprouts can withstand frost better than most vegetables, but are very sensitive to heat. Each plant is quite productive, providing 75–100 sprouts. As a member of the cabbage family, it requires substantial fertilizer.

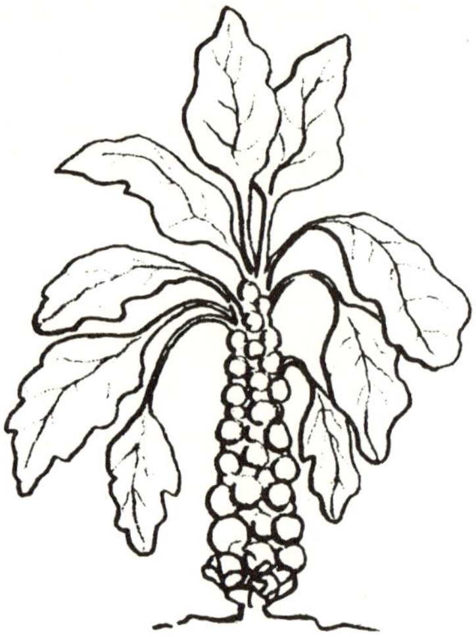

BRUSSELS SPROUTS

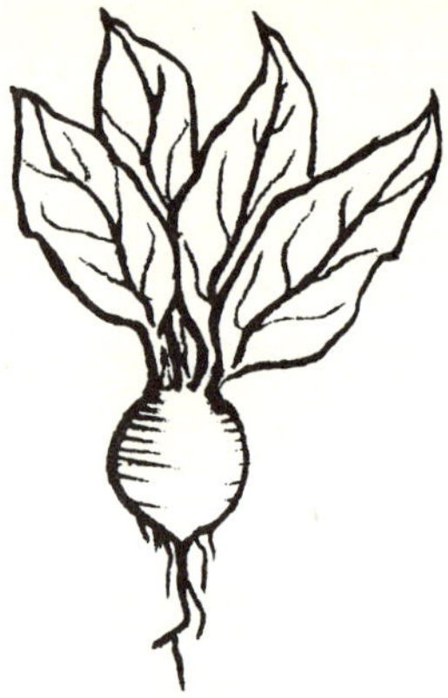

BEET

BROCCOLI

Characteristics of Different Vegetables

6. Cabbage

There are many types of cabbage—green, red, savoy, Chinese. A considerable amount of space is required for the amount of food produced. Cabbage is susceptible to many insects and diseases. Because of this, the planting site for cabbage should be changed so that the same soil is not used for cabbage more than once in a three year period. Substantial fertilizer is required.

7. Carrots

Carrots are easy to grow and tolerant of a range of temperatures, once the seeds sprout. Seed germination, however, can be a problem. The soil must be kept moist. Carrots have a high vitamin content.

CABBAGE

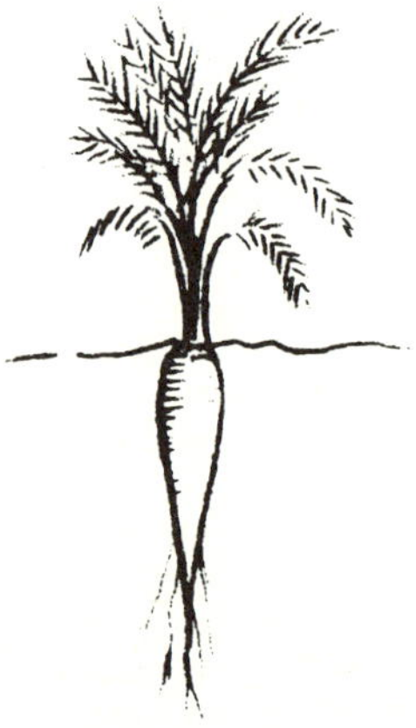

CARROT

Characteristics of Different Vegetables

8. Cauliflower

Cauliflower is more easily damaged by frost than other members of the cabbage family. Proper growth depends on a steady, cool growing season. The heads are "blanched" to make them white. This is accomplished by tying some leaves together to cover the head.

CAULIFLOWER

9. Celery

Celery requires attention. Regular watering and fertilizing are necessary over a long growing period (four months). The weather must be cool and steady during this period. All parts of the celery plant—tops, stalks and hearts—can be used as food or relish.

CELERY

Characteristics of Different Vegetables

10. Chard

Swiss chard is a good substitute for spinach. It is easier to grow and can withstand higher summer temperatures. Succession planting is not necessary as each plant produces continually. The leaves are cooked like spinach. The stalks are cooked like asparagus.

CHARD

11. Corn

There are many varieties of corn—tall, short, white, yellow, popcorn, early, late. Corn takes up a lot of space for 2½ to 3 months. However, small plants such as radishes, lettuce, etc. can be planted among the corn. Substantial amounts of water and fertilizer are required. To get a continuous crop, succession planting is used, or different varieties (early, middle, late) are planted simultaneously. Corn is pollinated by the wind, and therefore should be set out in multiple rows or blocks rather than a single row.

CORN

Characteristics of Different Vegetables

12. Cucumbers

Cucumbers tend to cover a wide area if left alone. Therefore, they are generally trained to grow vertically up supports. The plants must be tied to the supports, as they do not cling. Cucumbers are very sensitive to frost. Frequent watering is needed to keep the soil moist. The plants are subject to mildew if the summers are foggy and damp.

13. Eggplant

Eggplant is a large plant which is quite decorative. Substantial amounts of fertilizer and water are required. To reduce the threat of insects and disease, the planting site should be changed so that the same soil is not used for eggplant more than once in a three year period.

CUCUMBERS

EGGPLANT

Characteristics of Different Vegetables

14. Lettuce

Lettuce is easy to grow and can be planted among other larger vegetables. Or it can be planted in space to be occupied later by warm season vegetables, such as corn. Lettuce does poorly in hot weather, and likes a temperature of 50°–60°F.

The two main types of lettuce are head lettuce and leaf lettuce. Leaf lettuce grows faster, and is harvested continually, as the leaves regrow, over a period of time. For head lettuce, the entire plant is pulled at once. Leaf lettuce is better for warm climates.

Lettuce is easy to grow from seeds, but the seeds may not sprout if the weather is too warm. Frequent watering and shade may be needed if the weather becomes hot.

LETTUCE

HEAD LETTUCE

LEAF LETTUCE

Characteristics of Different Vegetables

15. Melons

The main types of melons are cantaloupe (also called muskmelon) and watermelon. Melons are hot weather plants which require a long warm growing season and a lot of space. Cantaloupes and other smaller melons can be grown vertically on supports.

Melons need substantial quantities of fertilizer and a lot of water. The melons may have a bitter taste if the weather is cool.

MELONS

CANTALOUPE

WATERMELON

16. Onions

Onions are a cool season vegetable but can tolerate a fairly wide range of temperatures. There are two most popular types: bulb onions and green onions (scallions). Bulb onions can also be harvested as small green onions a few weeks after planting.

Onions can be grown from seeds, seedlings, or "sets" (small bulbs). Sets are the most common method of growing bulb onions, but the available varieties are limited. Green onions are usually grown from seed.

ONIONS

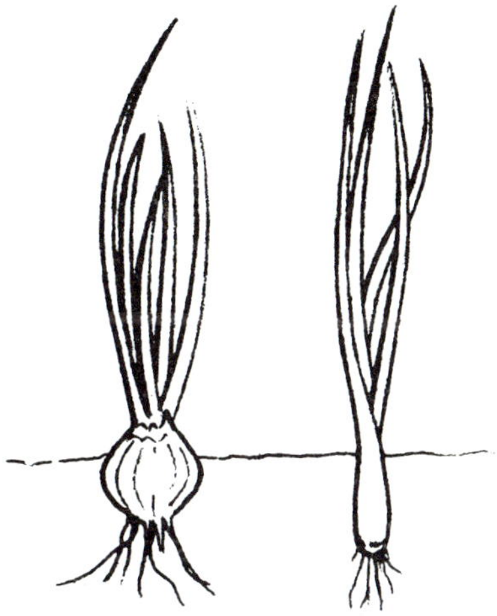

BULB GREEN

Characteristics of Different Vegetables

17. Peas

Peas are a cool-season crop which grow well in cool, moist air and soil. If conditions are right, the plants will produce a considerable amount of food in a short time.

Like beans, peas can be grown as bushes or vines. If grown as vines, they will need support.

Pea seeds may be wrinkled or smooth. Most varieties are wrinkled. Seeds are usually "inoculated", which is a means of providing special soil bacteria. This is easily applied before planting the seeds.

Some of the pea seeds may rot in moist soil. This problem can be reduced by ensuring good drainage (see Preparing the Soil, Page 21).

PEAS

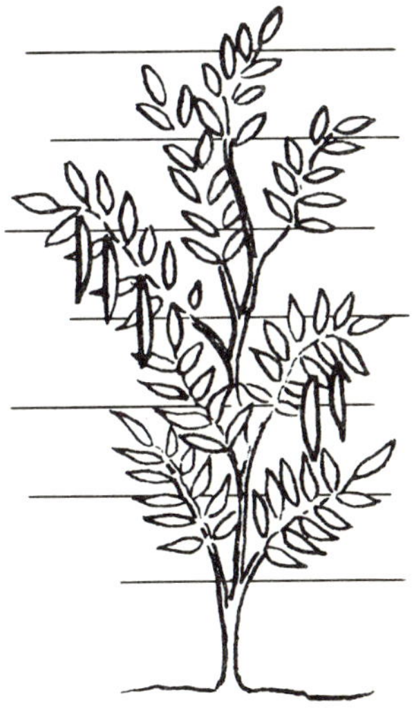

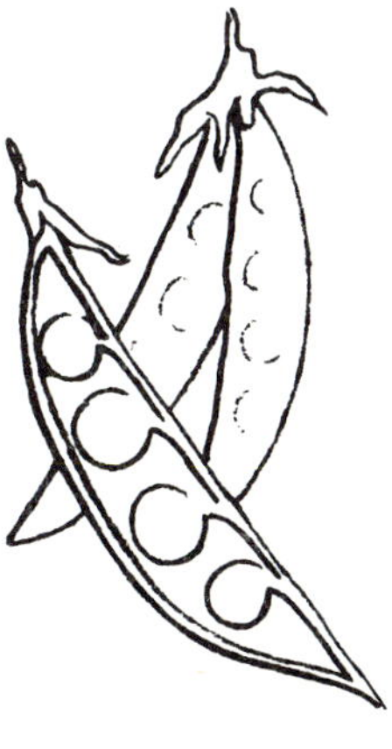

Characteristics of Different Vegetables

18. Peppers

There are many different kinds of peppers, but the two major types are sweet (or bell) and hot. Pepper plants are attractive and make good ornamentals. Large amounts of fertilizer and water are needed.

Of the two types, sweet peppers are adaptable to a wider range of climatic conditions. Hot peppers require long, warm summers.

SWEET PEPPERS

HOT PEPPERS

19. Potatoes

Potatoes are a cool season vegetable and require a large amount of space for a good length of time. Precut seed potatoes ("sets") are planted. Sprouts grow from the eyes. Potatoes bought from markets often have been treated with a sprout inhibitor and will not grow properly if planted.

Potatoes are subject to many insects and diseases.

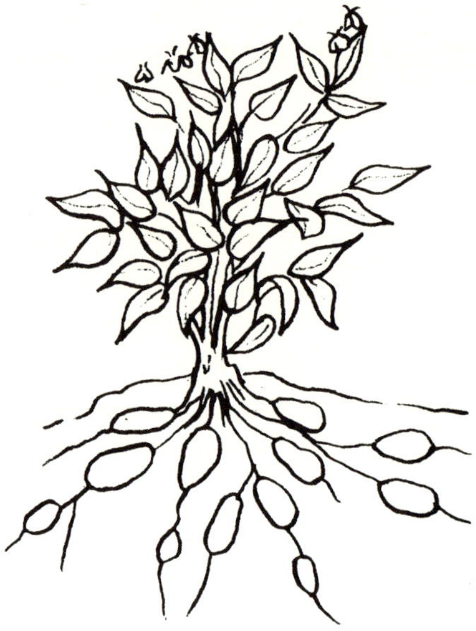

POTATOES

Characteristics of Different Vegetables

20. Radishes

Radishes are one of the easiest vegetables to grow. They grow rapidly and require little space. This makes it easy to plant them in spaces between other, larger vegetables. Since they are a cool season crop, they can be planted in areas which are later to be planted with warm season vegetables such as corn and tomatoes. In addition to ordinary radishes, there is a winter type which takes longer to grow, is larger, and must be peeled before eating.

RADISHES

21. Spinach

Spinach is a cool weather vegetable which can be grown before and after warm weather vegetables. In harvesting, the center sprout is left to form new leaves as the old ones are cut.

Spinach is grown best from seeds, planted at one to two week intervals. Spinach grows well if fertilized heavily.

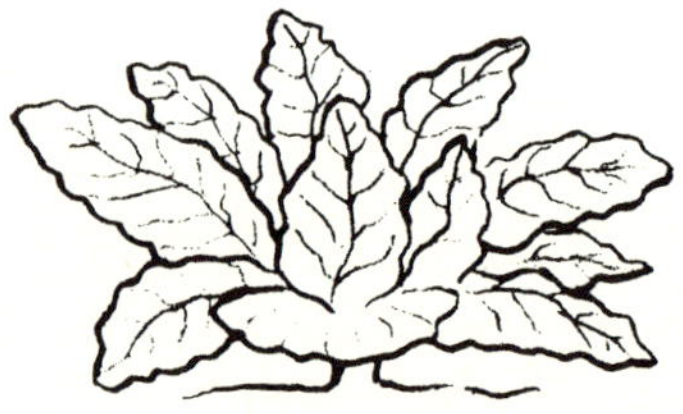

SPINACH

Characteristics of Different Vegetables

22. Squash

Although there are many kinds of squash, there are two main types, summer squash and winter squash. Summer squash is smaller and grows more quickly than winter squash. Summer squash is harvested while still young and the skin is soft, and the entire squash is eaten. Winter squash has a hard skin which is not eaten.

Squash generally requires a large growing space. However, summer squash can produce a large crop from only a few bushes or vines. It is sometimes grown vertically on supports. In this case both the plant and squash must be tied up.

Both types of squash need to be watered and fertilized heavily.

SQUASH

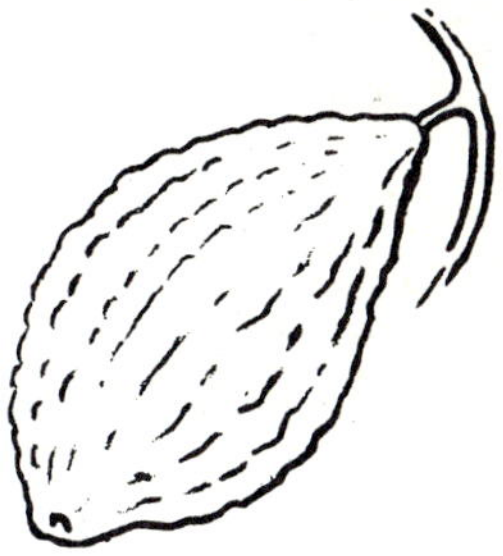

WINTER SQUASH

SUMMER SQUASH

Characteristics of Different Vegetables

23. Tomatoes

Tomatoes are a very popular vegetable and are easy to grow if there is warm weather. Tomatoes are usually grown from seedlings, and are supported by stakes or wire fencing. If stakes are used, the plants must be tied, as there are no climbing tendrils. Since the plants keep producing once started, there is no need for succession planting.

The planting site should be changed from year to year to avoid soil-borne plant diseases.

24. Turnips

Turnips are a cool weather vegetable which, like beets, can produce a large crop in a relatively small area. Also like beets, both their tops and roots can be eaten. Turnips grow rapidly, and need lots of water.

TOMATOES

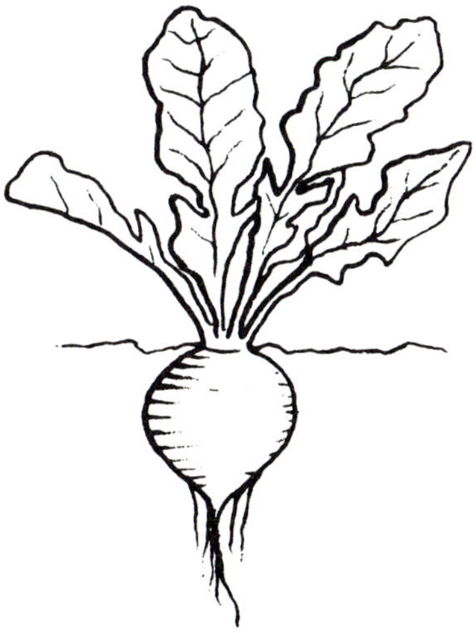

TURNIP

PLANNING THE GARDEN

Characteristics of Different Vegetables

Compatibility of Vegetables

There are many different opinions among gardeners as to which vegetables should be planted together and which should not. Certain vegetables seem to do well in the presence of other vegetables, while some seem to restrict the growth of others. Unfortunately, there is relatively little factual information available.

An established fact is that the "legumes", beans and peas, grow well with all vegetables except onions. Beans and peas are valuable also from the fact that they put back nutrients into the soil. Other vegetables take them out.

Onions, however, do grow well with the cabbage family (broccoli, brussels sprouts, cabbage and cauliflower), with beets, and with tomatoes. Corn, cucumbers, melons and squash grow well together, as do lettuce and radishes.

Some vegetables are "light feeders", and remove small amounts of nutrients from the soil. Others are "heavy feeders" and remove large amounts. Beans and peas, as mentioned above, put nutrients back in. By changing the position of these three classes of vegetables in the garden, soil fertility can be maintained or even improved.

The order of planting to follow is this: heavy feeders should be followed by beans and peas, which replace the lost nutrients. Then the light feeders are planted. Of course, you can begin anywhere in the cycle.

It will not always be convenient to follow this sequence. Nevertheless, it is a desirable feature to keep in mind in your planning.

LIGHT FEEDERS

beets, carrots, onions, radishes, turnips

↓

HEAVY FEEDERS

broccoli, brussels sprouts, cabbage, cauliflower, celery, chard, corn, cucumbers, eggplant, lettuce, melons, peppers, potatoes, spinach, squash, tomatoes

↓

NUTRIENT REPLACERS

beans, peas

▶ Location and Size of Garden

The planner is usually faced with several possible choices for the location of the vegetable garden. The major considerations are sunlight, convenience, and a few other factors to be described.

The amount of sunlight is important, since all vegetables, even cool season types, need sun. If possible, locate your garden in an area which receives six hours or more of sunlight per day. The shady side of buildings (the north and east sides), should generally be avoided. Also, avoid the shade from large trees, high hedges, etc.

The vegetable garden should be within a convenient distance of the house. This will make it easy to watch the progress, and harvest at the right time. Animals will cause less trouble. A water supply will normally be handy near the house.

The garden should be located away from trees. In addition to the shade problem mentioned, tree roots will compete with the vegetables for soil moisture and nutrients. Tree roots may typically extend out to the "drip line", so the area within this line should be avoided.

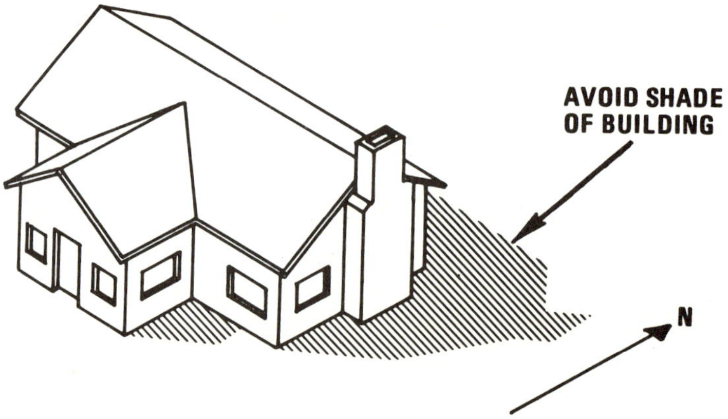

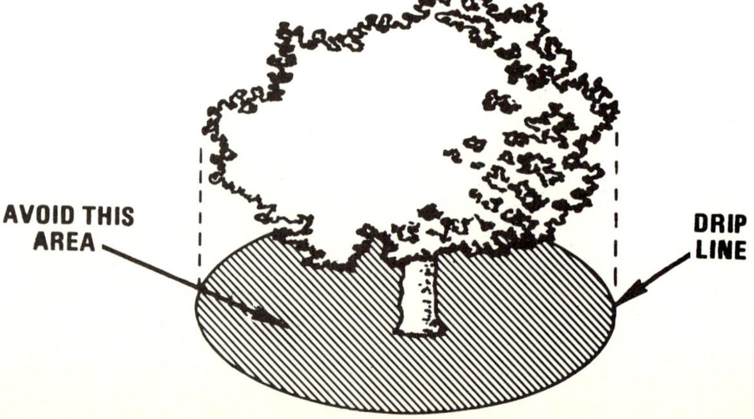

Location and Size of Garden

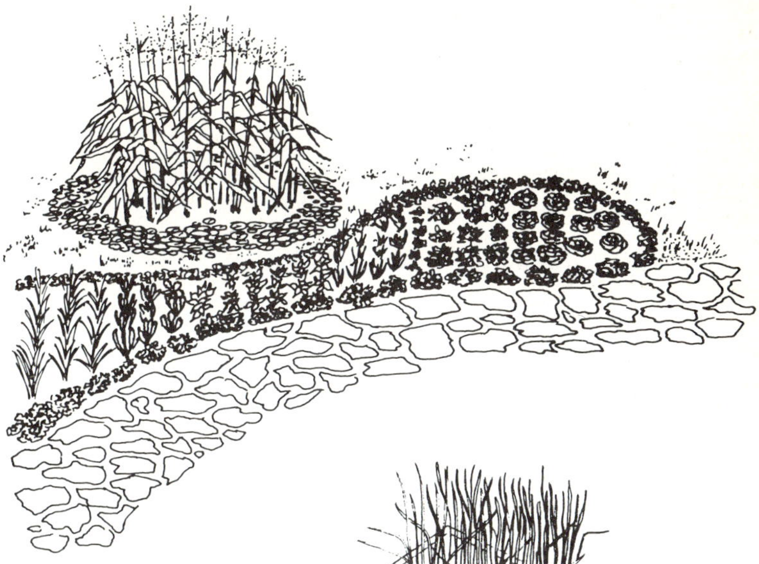

If there is a choice of locations, choose an area which has good soil and good drainage. Avoid low spots where water collects. As described later, the soil will be reworked to a considerable extent. But you will have that much better start if you can find an area with deep loose top soil rather than gravel, rock, or hardpan.

The garden can be located in more than one spot and can be any convenient shape. One practice that is quite common is to grow some vegetables right in with the flowers. Or they can be in borders grown along walls or fences. Many vegetables, as described in a later section (Growing Vegetables in Containers, Page 45), can be grown in containers on the patio or deck. Any of these methods can be used instead of, or in addition to, the conventional rectangular garden.

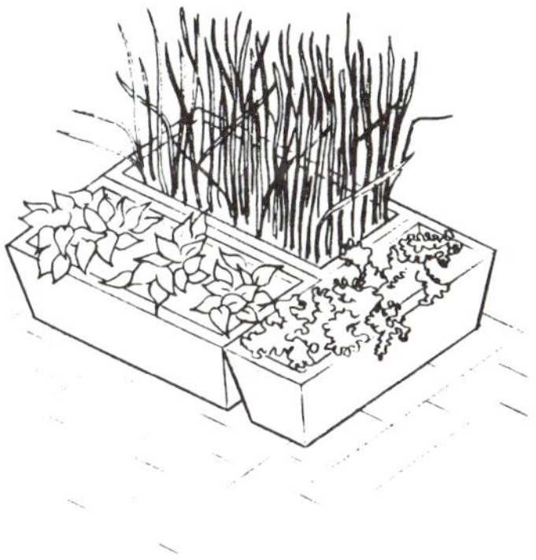

Most beginning vegetable gardeners start with too large a garden. It is a much better idea to begin with a small plot and expand as you become more familiar with gardening. If you start with a small plot you will tend to pay more attention to preparing your soil properly and the individual plants will get more attention. If you start with a large garden, you may find you don't have time to give the garden adequate care. The results may be unnecessarily discouraging.

Location and Size of Garden

You may plan to preserve some of your vegetables, as described in a later section of this book (Preserving Vegetables, Page 55). In this case, you will eventually want a larger garden than if you immediately eat your vegetables. Even so, it will be better to work your way into a larger garden gradually. Start with a small area and add other small areas over a period of time.

Different kinds of vegetables occupy different amounts of space in the growing process. This may influence your choice of vegetables, depending on space limitations. In any event, it will affect the amount of space allotted to each vegetable in laying out the garden. Also, the heights of different plants will affect their location in the garden: taller plants are placed toward the north in order to reduce shadows for the smaller plants.

For planning purposes, vegetables fall into three size groups, shown at the right. Melons and winter squash are in a class by themselves. They require lots of room.

The second group includes many other vegetables which are quite large when fully grown.

The third group includes the smallest vegetables. These may often be "intercropped" with the larger vegetables (as described later, Page 20).

VEGETABLES GROUPED BY SIZE

Group 1 (Largest)	
Melons	Winter Squash
Group 2 (Medium to Large)	
Asparagus	Cucumbers
Beans	Eggplant
Broccoli	Peas
Brussels Sprouts	Peppers
Cabbage	Potatoes
Cauliflower	Summer Squash
Corn	Tomatoes
Group 3 (Smallest)	
Beets	Onions
Carrots	Radishes
Celery	Spinach
Chard	Turnips
Lettuce	

PLANNING THE GARDEN

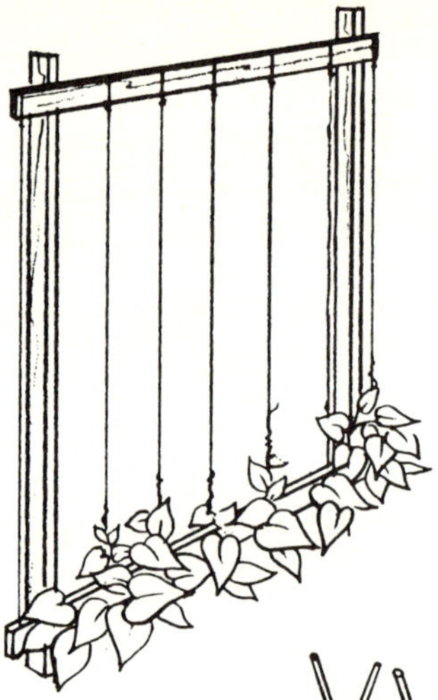

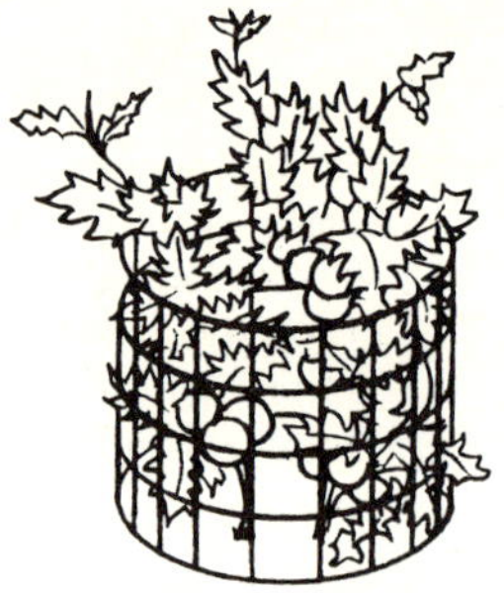

Location and Size of Garden

Some of the vegetables can be grown on supports such as poles, trellises, etc. Instead of spreading out, the vegetables grow up the supports. Since you can place the plants closer together, you can grow more vegetables in a given area.

Vegetables which can be grown in this manner include beans, cucumbers, eggplant, melons, peas, peppers, squash, and tomatoes.

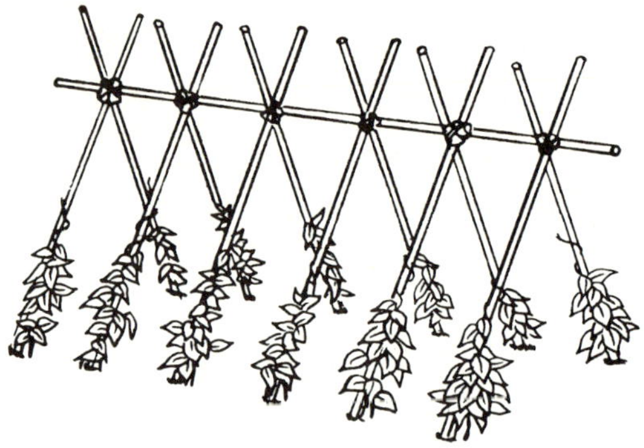

▶ Laying Out Your Garden

To begin laying out your own garden, a pencil and paper are useful. Start making a sketch as ideas occur to you and as you review the following points.

Your garden, of course, can be any size and shape that will fit the space available. But, as suggested earlier, it is wise to start with a small area and expand later. A surprising amount of vegetables can be grown in a small area, if they are given proper care—which is more likely if the area is small. This will become more clear when we discuss Preparing the Soil, Page 21.

Recall first that the garden should be placed in a sunny location, near the house if possible. Low spots where water accumulates should be avoided.

For planning purposes, a useful small area, or "module", is 4 feet by 4 feet. We can make the garden as large as we like by adding other 4 foot square modules.

There are several reasons for choosing a module, 4 feet on a side. First, each row will be limited to a length of 4 feet, which tends to limit the amount planted. Additional rows can always be added as needed.

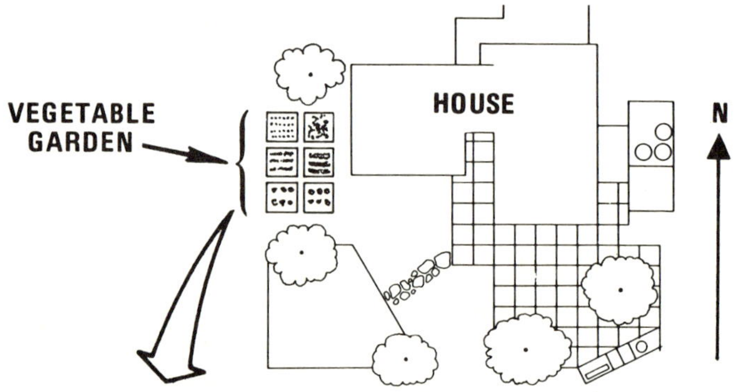

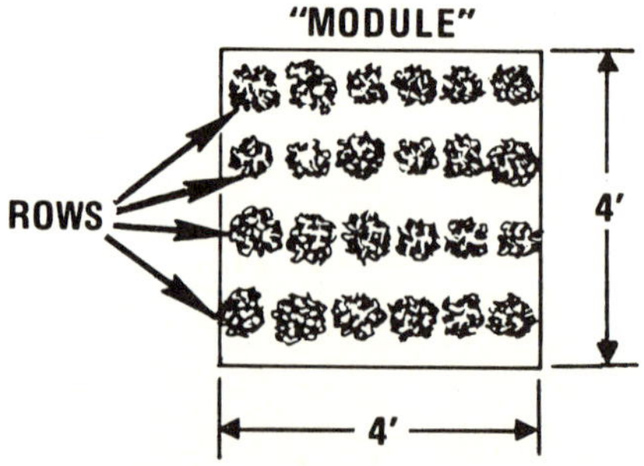

Second, a size of 4 feet means that all vegetables will be within easy reach from any side. This is helpful both in caring for the vegetables and in harvesting.

Laying Out Your Garden

A third reason for choosing the 4' x 4' module is as follows: A gross measure is needed for initial planning of space requirements. The 4' x 4' module serves that purpose. It provides an approximate measure of the space needed for a small family (2 or 3 people) for a single vegetable. That is, some vegetables require more space, some less, but a 4' x 4' plot is about the average space requirement of the 24 listed vegetables.

This 4' x 4' module, then, is a useful tool to use in laying out the garden. The accompanying table provides a rough indication of the number of such modules which would be normally used for each vegetable. These numbers are in no way meant to be exact. They are gross approximations of a small family's needs.

You may find another module size more useful—for example, 4' x 6', 4' x 8', or some other size. Choose whatever size you wish. For the above three reasons, you will find the modular approach useful.

VEGETABLE NEEDS FOR SMALL FAMILY

Vegetable	Approx. No. of Modules	Vegetable	Approx. No. of Modules
Asparagus	1	Eggplant	1
Beans	1	Lettuce	½
Beets	½	Melons	2
Broccoli	1	Onions	½
Brussels sprouts	1	Peas	1
Cabbage	1	Peppers	1
Carrots	½	Potatoes	2
Cauliflower	1	Radishes	½
Celery	½	Spinach	½
Chard	½	Squash	1–S, 2–W*
Corn	2	Tomatoes	1
Cucumbers	1	Turnips	½
		Average:	1

*S — Summer squash
W — Winter squash

Laying Out Your Garden

Seeds or seedlings (small, young plants) are planted in 4-foot rows. The distance to allow between rows, and the distance between seeds or seedlings within a row, depends on the individual vegetable, as described later (Page 29).

Large plants such as cauliflower or squash require 2–3 feet between rows. Small plants like carrots and radishes require only about 8 inches. Distance between seeds in a row may be as little as 1/2 inch or as much as 3 inches. It is these differences in space requirements that lead to differences in the approximate number of modules shown in the preceding table.

The module is easier to use for planning purposes than such measures as "number of feet of row". Because of the above considerations, this measure requires additional calculation. But differences in climate, soil, appetites, etc. do not warrant such detailed calculations.

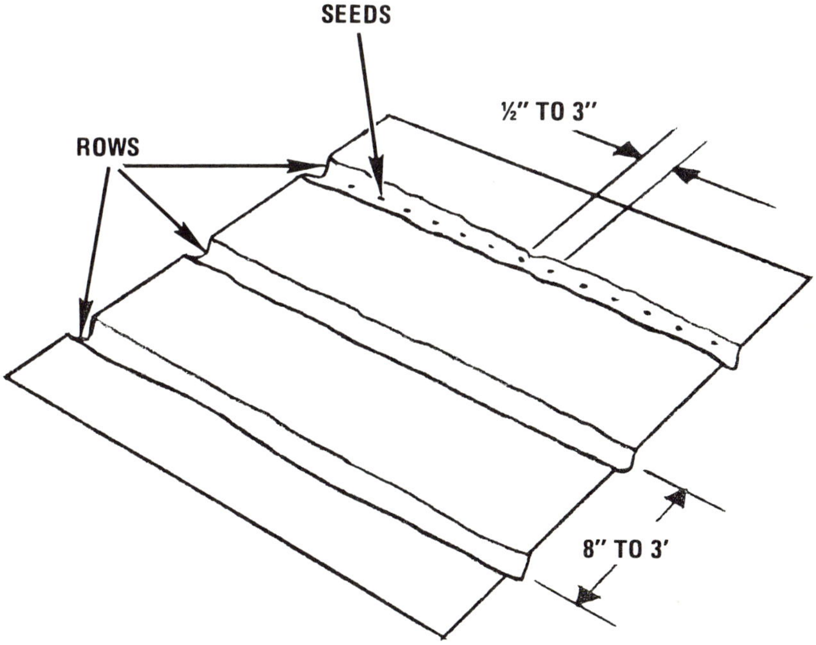

PLANNING THE GARDEN

Laying Out Your Garden

Before describing how to use the modules in laying out a garden, the general position of the different vegetables needs to be discussed.

The position of vegetables in a garden will depend mainly on the size of the plants. Tall plants, particularly those which are supported, should be at the rear and away from the sun. The north side is best. Then their shadows will not deprive smaller plants from needed sunlight. Tall plants which should be at the north end of the garden are as follows: pole beans, corn, cucumbers, peas, and tomatoes.

Next in line, toward the south, should be placed the medium to large plants which are not usually supported: bush beans, broccoli, brussels sprouts, cabbage, cauliflower, eggplant, melons, peppers, potatoes, and squash.

The rest of the garden is preferably south of the larger plants, and is taken up with smaller vegetables. These include: beets, carrots, celery, chard, lettuce, onions, radishes, spinach, and turnips.

One vegetable, asparagus, has not been mentioned in the above three lists. Asparagus is unique in that it is the only perennial on the list. It is planted once only, and produces for years. Because asparagus is a perennial, it should occupy a separate spot altogether, out of the way of other vegetables. It is a tall plant, so should not be placed where its shadow will fall on other vegetables.

GENERAL POSITION OF VEGETABLES IN GARDEN

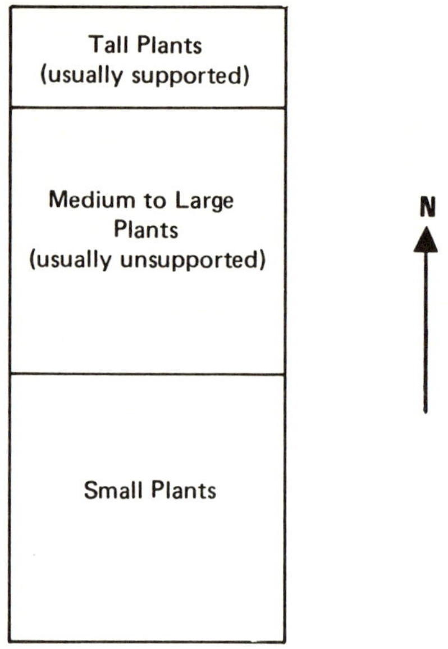

Laying Out Your Garden

The garden can be developed one module at a time. You can stop after one module, two, three, or however many you want. For illustrative purposes, shown here are six modules. We will first consider a cool season garden, and then a warm season one.

In the nothernmost part of the garden, supported on a fence or other means of support, are peas. Asparagus is also a candidate for this area, but it is assumed that asparagus, being a perennial, has been assigned to a different part of the yard.

The two north modules are divided in half, and the peas are in the northern half. This is because they normally require one module, as seen from the table (Page 17). Being tall and supported, this one module (total of two halves) needs to be in the northernmost part of the garden.

Moving south from the peas, the areas shown can include any combination of a number of vegetables: broccoli, brussels sprouts, cabbage, cauliflower, or potatoes. These all occupy a fair amount of space, particularly potatoes.

The two southernmost modules can include any of the following smaller vegetables: beets, carrots, celery, chard, lettuce, onions, radishes, spinach, turnips. Any two of these vegetables can be planted in a single module, for a total of four in the two southernmost modules.

For some of these vegetables, succession planting and/or intercropping can be used (see Page 20).

COOL SEASON GARDEN

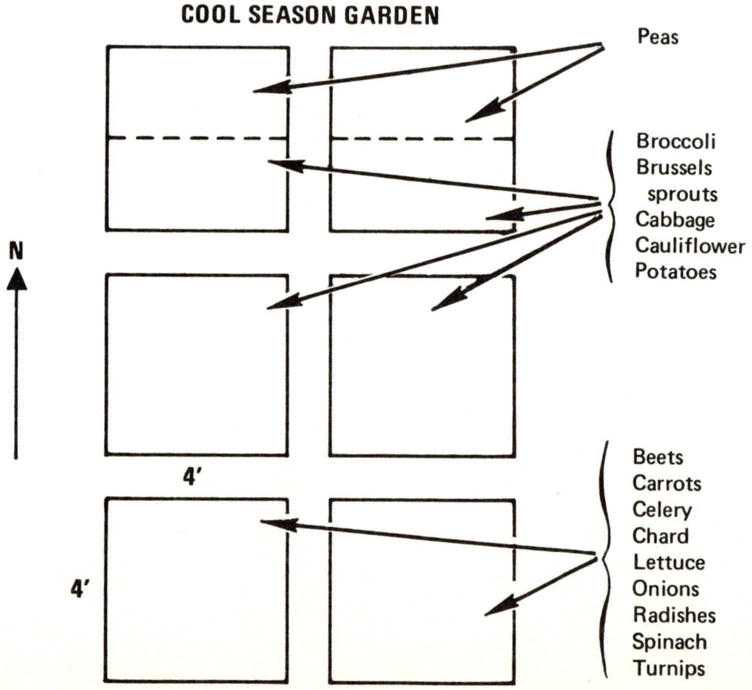

Laying Out Your Garden

In the warm season garden, pole beans, cucumbers, and tomatoes are the choices for the northernmost part of the garden. Like peas, they need support. Corn is also shown as a possibility. Since corn needs more room, both entire modules are planted. Succession plantings may be used for corn (Page 20).

All the space to the south can include bush beans, eggplant, melons, peppers, and/or squash. If the melons or squash are supported to conserve space, they should be planted towards the north near the other taller plants.

In most areas of the United States, the cool season garden is planted in spring, and the warm season garden replaces it later. Then again, still later, the cool season garden can replace the warm season garden in the fall. In some areas, where the summers are cool, the cool season garden may be used throughout the growing period.

WARM SEASON GARDEN

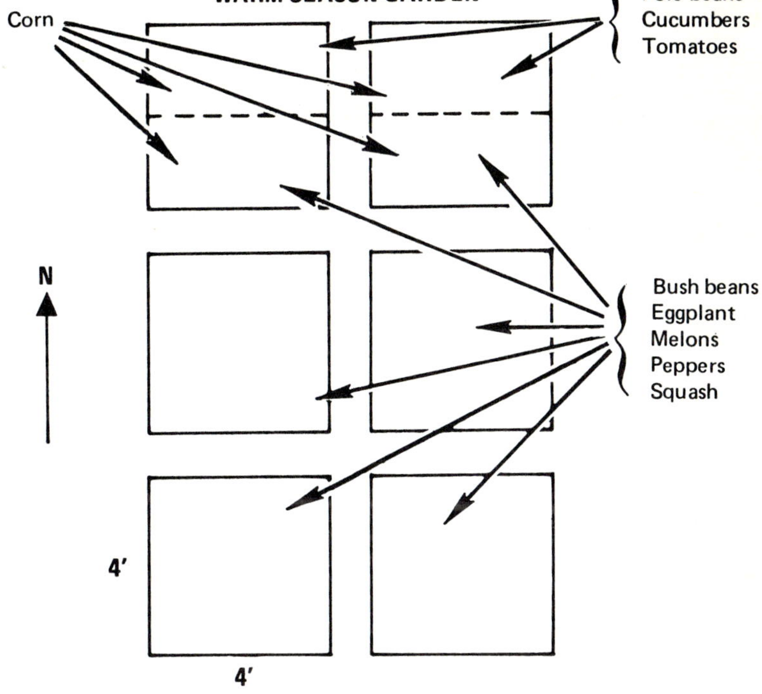

Laying Out Your Garden

Selecting Vegetables

Vegetables can be selected with the help of the information in the table on Page 6 and according to your own personal likes. The most important factor is to match the vegetables' growing season requirements to the climate in your area. Other factors which may influence your choice are requirements for space, support, and care.

If you intend to preserve vegetables by freezing, canning, or storing, you may wish to choose vegetables for which special varieties are suitable for this purpose. These varieties are given later in Preserving Vegetables, Page 55. The vegetables for which special varieties are available for preserving are shown at the right.

VEGETABLES FOR WHICH SPECIAL VARIETIES ARE AVAILABLE FOR FREEZING, CANNING, OR STORING

Freezing	Canning	Storing
Asparagus	Beans	Beets
Beans	Beets	Cabbage
Beets	Carrots	Carrots
Broccoli	Corn	Celery
Brussels sprouts	Peas	Onions, bulb
Carrots	Peppers	Potatoes
Cauliflower	Spinach	Radishes
Chard	Squash, winter	Squash, winter
Corn	Tomatoes	Turnips
Peas		
Peppers, sweet		
Spinach		
Squash		

Laying Out Your Garden

In addition, to help you in your selection, you should be aware of "succession planting" and "intercropping". These techniques are used to an advantage for many vegetables.

Succession planting of same crop: With many vegetables, it is better to plant only part of the allowed area at one time, rather than plant it all at once. This will spread out the harvest period to a more useful length. In this method, area 1 is planted first. Then, 1–2 weeks later, area 2 is planted, and 1–2 weeks after that, area 3 is planted. Later, after area 1 is harvested, it can be replanted if there is still enough time left in the growing period. Similarly for areas 2 and 3.

Vegetables for which this practice is common include: beans, beets, broccoli, cabbage, carrots, celery, corn, lettuce, peas, radishes, spinach, and turnips.

SUCCESSION PLANTING OF SAME CROP

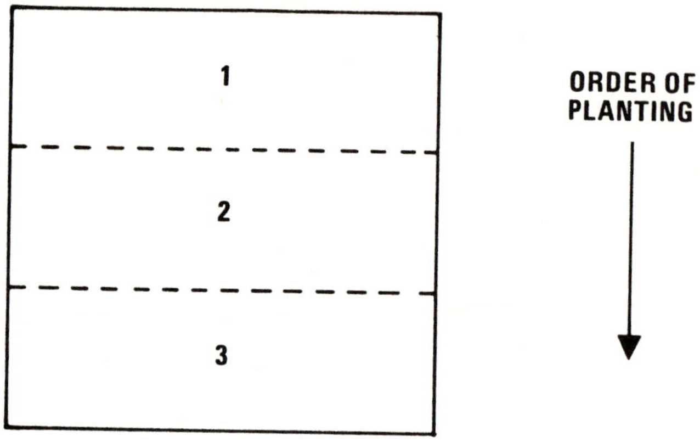

Laying Out Your Garden

Succession planting of different crops: To make the best use of garden space, it is common to plant cool season and warm season plants in the same space at different times. For example, the place used for peas, a cool season vegetable, may later be used for peppers, a warm season vegetable. Or broccoli (cool season) may be followed by summer squash (warm season).

Then, still later, the warm season vegetable can be followed by a cool season vegetable if desired, and if climatic factors permit.

Intercropping: Vegetables which are small and mature quickly can be planted among larger, slower-maturing vegetables without disturbing the latter. For example, leaf lettuce, green onions or radishes can be planted among corn or tomatoes and harvested before the corn and tomato plants begin bearing. Vegetables which are sometimes intercropped among slower plants include: beets, carrots, celery, chard, lettuce, onions, radishes, spinach and turnips.

SUCCESSION PLANTING OF DIFFERENT CROPS

INTERCROPPING

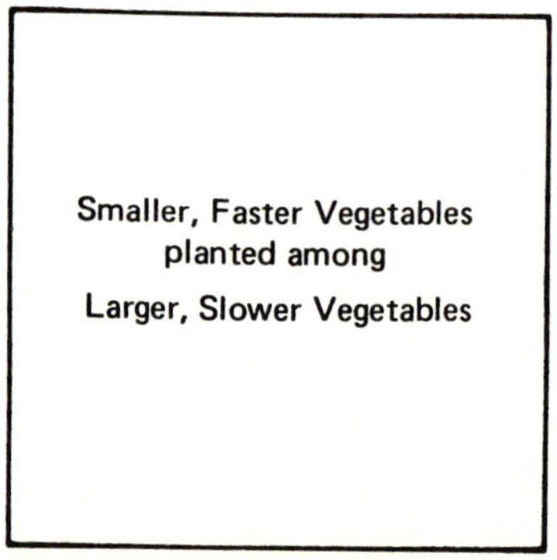

▶ Soil Characteristics

In contrast to the climate, the condition of the soil in the vegetable garden is to a large extent under the control of the gardener. Initially, the soil in one yard may differ drastically from that in another. But what that basically means is that more work will be required to develop proper soil in one area than in another.

The soil characteristics of most importance to the home gardener are the particle size and the amount of humus (decomposed organic matter). Particle size can vary from large sand particles to minute clay particles. Silt particles are between these two extremes in size.

If the soil is too sandy, water will pass right through and not be retained for vegetable plants to use. On the other hand, if there is too much clay, the particles tend to pack together too tightly and do not admit air or water evenly. When clay dries, it can become hard as a rock.

An abundant amount of decomposed organic matter, or humus, is the key to having a good soil to grow vegetables. Whether your soil is sandy or clay, or somewhere in between, it can be greatly improved by adding large amounts of humus.

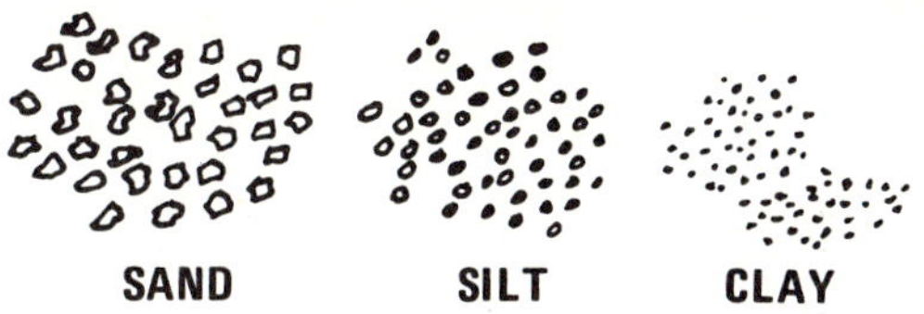

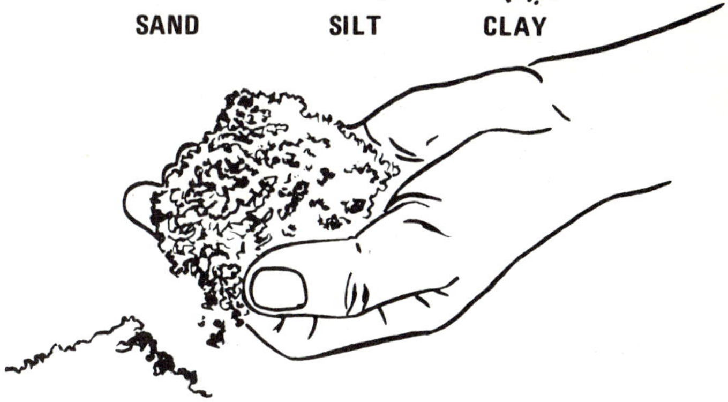

FRIABLE SOIL WITH HUMUS

Humus when mixed with the soil provides the right amount of air and absorbs the right amount of water for growing plants. The soil becomes light and crumbly—such soil is called "friable"

One of the best sources of humus is compost. Compost can be made easily from any available organic material, as described later (Page 24).

Soil Characteristics

The pH factor: The acidity or alkalinity of soil is measured by the "pH" (for potential of hydrogen). The pH is a number which varies from 0 to 14, with an average or neutral value of 7. If the value is less than 7, the soil is acid (sour); if greater than 7, the soil is alkaline (sweet).

Vegetables grow best in soils which are slightly acid, with a pH value between 6 and 7. When the pH is in this range, nutrients (or food for the plants) in the soil are available to the plants, and the plants absorb them. When the pH is outside this range, nutrients can be right next to the roots and yet literally not be available to the plants or be over-abundant in harmful quantities. Adding further nutrients by means of fertilizers, under these conditions, will not help the plant.

The pH value of your soil can be measured by kits available on the market, or samples of your soil can be sent to the Cooperative Extension Service in the county or at the state university. While such measurements may prove useful in some cases, if your soil is prepared in the manner described in the next section (Page 22), your soil will usually have a pH near the proper value. This is, again, a reason for using lots of humus: the pH automatically tends toward the correct value for growing vegetables.

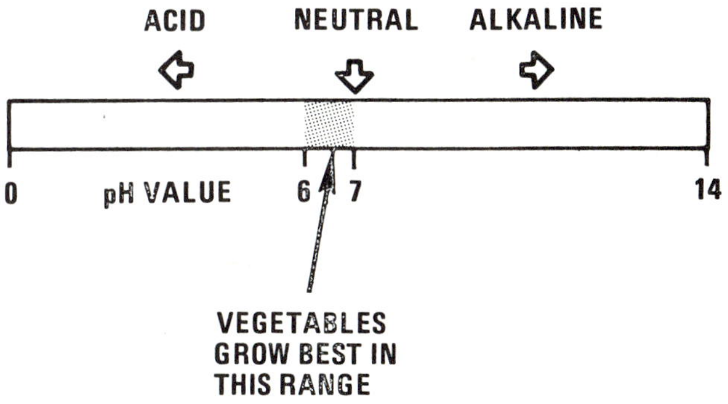

Soil Characteristics

N, P, and K: Many nutrients are needed by plants in order to grow properly, but the major ones are Nitrogen (N), Phosphorus (P), and Potassium (K). Large quantities of these three nutrients are consumed by all growing vegetables. Therefore, a continuing supply of nutrients is needed during the growing season.

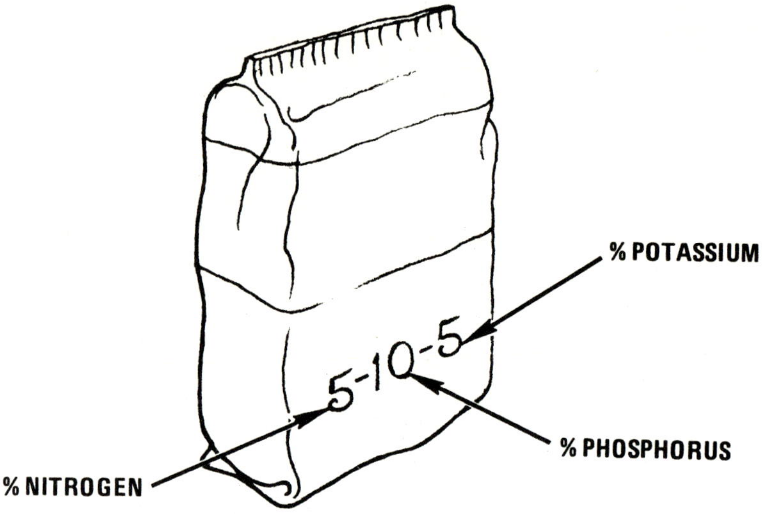

N, P, and K can be supplied by adding fertilizers—either organic fertilizers such as manure (or others discussed later), or inorganic commercial fertilizers. The latter may be conventional types or the timed-release type. Commercial fertilizers are always labeled—for example, 5-10-5, 10-20-10, etc. These numbers indicate the percentages of N, P, and K, respectively.

Earthworms: Earthworms are very valuable for improving the soil. They will tend to thrive and multiply in good soil for growing. They continually break up and mix the soil. You do not need to worry about adding earthworms. They will appear automatically in properly-prepared soil.

▶ Adding Soil Conditioners

The soil conditioners that are to be added include organic matter (humus) and N,P, and K. There are many different ways of adding these conditioners, depending on what is available to you. One method will be described, and some of the common substitutes that may be used will be indicated.

Procedures will be described in terms of a 4' x 4' "module", as described earlier. You can substitute another module size if you wish. As many modules as desired can be prepared.

Since preparing the soil raises the soil level [1] above that of the surrounding ground [2], it is convenient to put a wooden frame [3] around the module. However, while this will be described, it is not necessary.

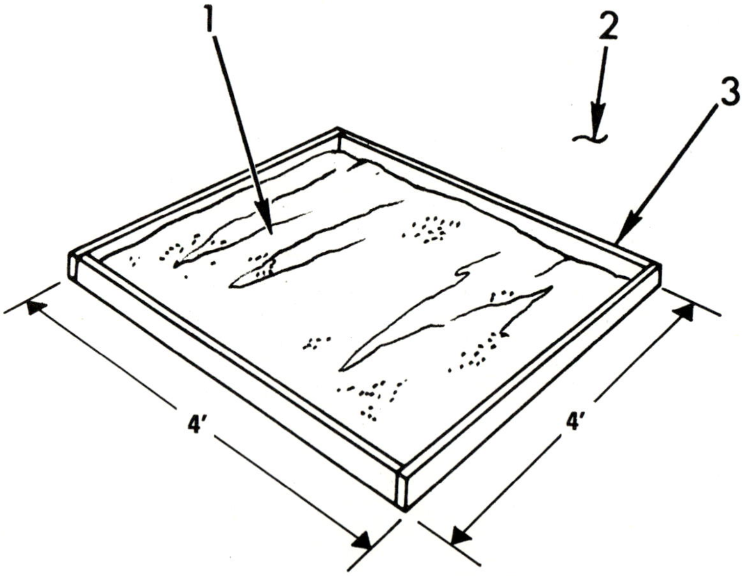

Adding Organic Matter

The organic matter that will be added is compost. A method of making compost is described later (Page 24), or it may often be purchased. If it is made, a time period of 3 weeks to 3 months should be allowed.

If compost is not available, a temporary substitute is manure. The manure should be several months old, not fresh. Horse manure is better than steer manure, as there is less salt.

Adding Soil Conditioners

Tools:

Tape measure
Spading fork (optional)
Spade
Rototiller (optional)
Rake

Supplies:

Compost or manure (preferably horse) – 5 cu ft

1. Measure 4' x 4' plot of ground in location chosen for vegetable garden. Using spade [3] mark edges [4].

2. Using spade [3] or spading fork [1], dig and turn earth in the plot to a depth of about one foot. If available, a rototiller [2] can be used. However, a rototiller will generally be useful only if you are using a larger module size than 4' x 4'.

3. Using spade [3] and rake [6], spread a 4-inch layer of compost [5] on the plot.

4. Using spade [3] or spading fork [1], turn compost [5] into soil. Remove any rocks.

5. Using rake [6], rake plot level.

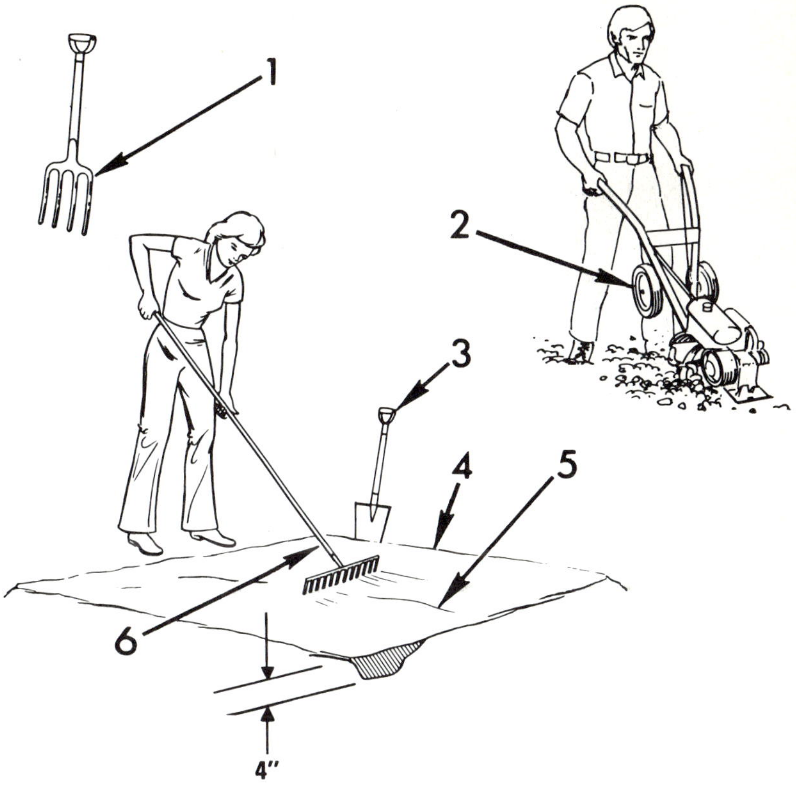

Adding Soil Conditioners

Adding N, P, and K

Nitrogen (N), phosphorus (P) and potassium (K) must now be added to the soil. N, P, and K can be added by adding manure (for N), superphosphate or bone meal (for P) and wood ashes or greensand (for K). Instead of these, a conventional commercial or a timed-release fertilizer such as "Precise" Timed Release Vegetable Food 9-12-6 may be used. The latter is preferred because it releases nutrients over a period of time, whenever the vegetables are watered.

Tools:

Spade
Rake

Supplies:

Manure (preferably horse) – 2 2/3 cu ft
Superphosphate or bone meal – 1 lb
Wood ashes or greensand – 1 lb
or
5-10-5 fertilizer – 1 lb
or
9-12-6 "Precise" Timed Release Vegetable Food – 4 oz

If using 5-10-5 or 9-12-6 fertilizer, go to Step 3.

1. Using spade and rake, spread a 2-inch layer of manure [1] on plot.

2. Sprinkle one pound of superphosphate [2] or bone meal and one pound of wood ashes or greensand on plot. Go to Step 4.

3. Spread one pound of 5-10-5 fertilizer or 4 ounces of timed-release fertilizer on plot.

4. Using rake, rake the materials into the top 6 inches of the soil. Rake smooth.

Adding Soil Conditioners

Wood Framing

Framing of the module is optional. Placing a wooden frame around the module keeps the soil in place and makes a neater appearance.

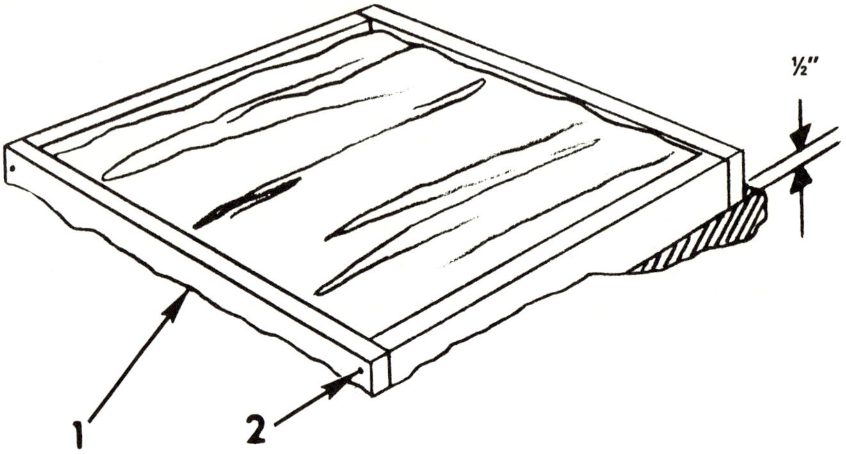

Tools:

Hammer

Supplies:

2" x 4" boards, 4' long (4)
Nails

1. Place 2" x 4" boards [1] around edge of module. Using hammer, tap into soil about 1/2 inch.

2. Using hammer, nail corners [2] for security.

Adding Soil Conditioners

Making Compost

Compost is the result when vegetation and other organic material decays. A compost pile speeds up the decay process which occurs in nature.

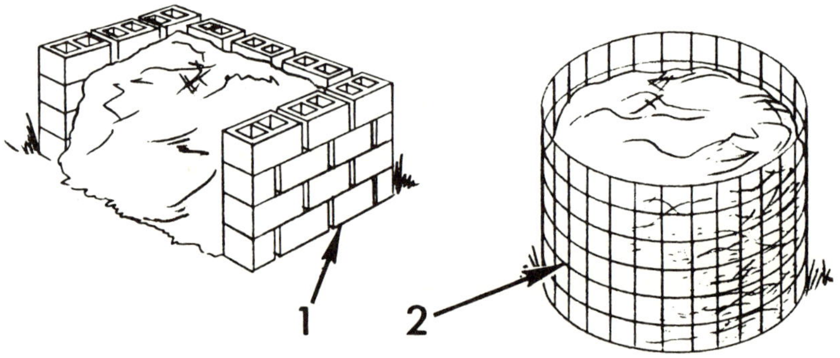

Compost making requires from a few weeks to a few months, depending on the method used. The standard, slower method will be described first. Then a means of shortening the process will be described.

If the instructions are followed, there should be no flies or offensive odors. The compost pile will be moist but not soggy, and well ventilated.

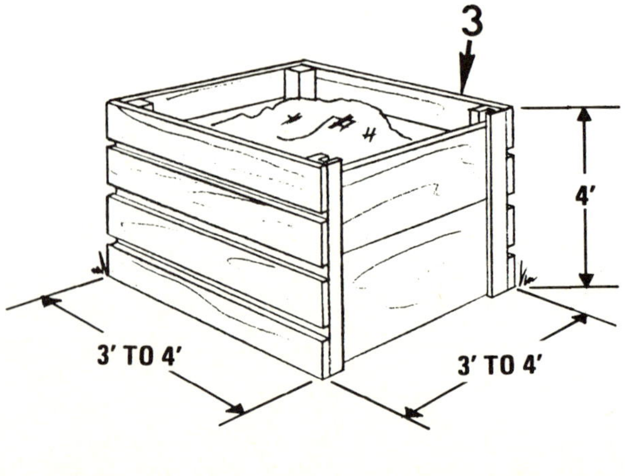

A compost pile can be made in the open, or, better, a simple bin can first be constructed. A bin prevents the pile from blowing or falling over. A typical bin is about 3 to 4 feet on a side by 4 feet high, although larger bins are common. The height is necessary to generate the required heat for proper composting.

The bin can be constructed from concrete blocks [1], wire fencing [2], wood [3], or any other similar materials. Leaving one face of the bin open or easily moveable, makes access easy.

Adding Soil Conditioners

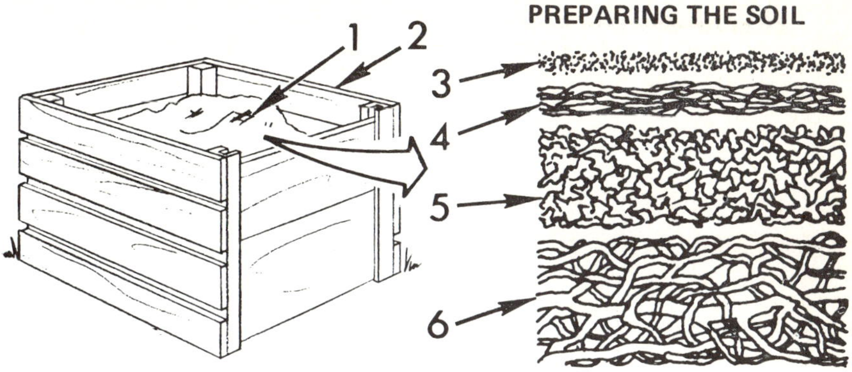

Tools:

Spade
Spading fork
Pole or stick

Supplies:

Waste vegetation: leaves, brush, grass clippings, twigs, branches, kitchen garbage (other than meat, fat, or bones)
Fresh manure, bone meal, blood meal, or cottonseed meal
A source of water

A layer of rough materials on the bottom of the pile allows air to circulate.

1. On the bare ground, or at the bottom of the bin [2], spread a layer of rough materials [6] such as twigs, brush, corn cobs, etc. The layer should be about 6 inches deep.
2. On top of the first layer, spread a layer of mixed vegetation materials [5] such as grass clippings, leaves, weeds, hay, kitchen garbage (excluding meat, bones, and fat), etc. The layer should be of mixed materials, not just one kind, and should be about 6 inches deep.
3. Using spade, spread a 2-inch layer of fresh manure [4] or a light sprinkling of a nitrogen substitute – bone meal, blood meal, or cottonseed meal.
4. Using spade, spread a 1-inch layer of soil [3].
5. Sprinkle water on the pile until moist, not soggy. Pile should be kept about as moist as a squeezed wet sponge.
6. Repeat Steps 2 through 5 until pile is about 5 feet high. The top layer [1] should be soil.

The size of the pile will shrink considerably with time. You can either add to the pile as it shrinks or make a new one, depending on when you will need the compost.

Adding Soil Conditioners

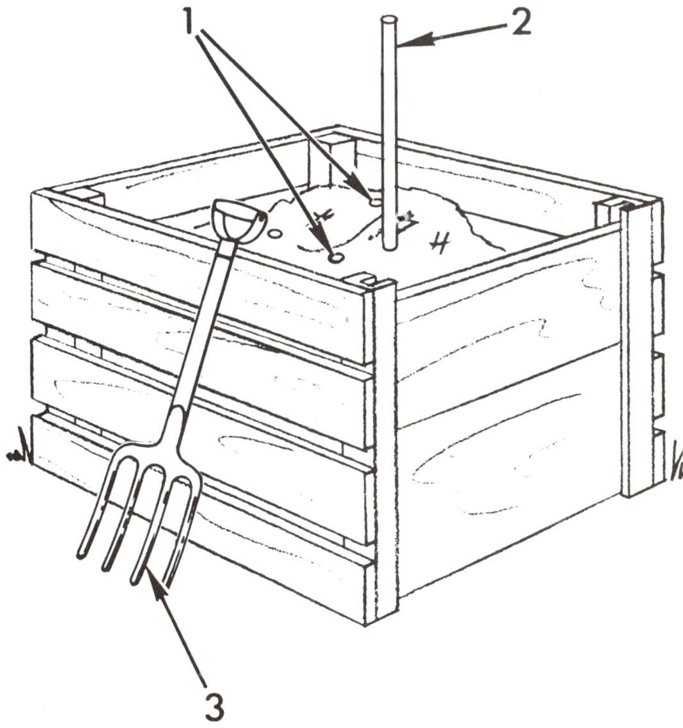

Bacterial action is helped by poking holes in the pile to provide air.

7. Using pole or stick [2], poke holes [1] vertically in top of pile, reaching to bottom of pile. Holes should be about 2 feet apart.
8. Keep pile moist but not wet by sprinkling with water about once a week. Pile should be about as moist as a squeezed wet sponge.
9. Using spading fork [3], turn pile every 2 to 3 weeks. Move outer material that is not decomposed into center of pile. Try not to disturb bottom layer of rough materials.

WARNING

Pile can become quite hot (130°–160°F). Check cautiously.

10. Check moisture and temperature periodically by reaching in hand. If pile is not heating up, add manure or substitute (bone meal, blood meal, cottonseed meal). If moisture is not proper, adjust watering.

Compost is ready to use in about 3 months. Material becomes brown and crumbly and has an earthy smell.

PREPARING THE SOIL

Adding Soil Conditioners

The composting process can be greatly speeded up if the vegetation and manure are first mixed and put through a shredder [1]. If a shredder is not available, a power lawn mower [2] may be used to cut up the materials, a small pile at a time. If this method is used, the pile should be turned after 3 to 4 days, and repeated every 3 to 4 days. The compost will be ready in 2 to 3 weeks.

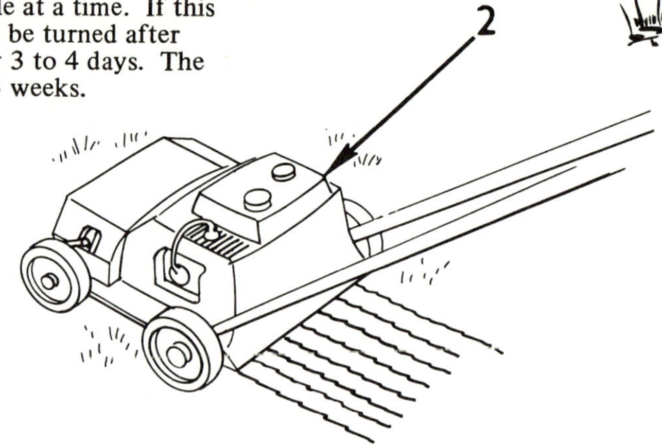

PLANTING

▶ When to Plant

Timing the initial planting of individual vegetables will depend primarily on whether the vegetable is a cool-season or warm-season crop.

Cool season vegetables (see accompanying table) tend to be the leafy and root vegetables. They grow well in cool weather. In order to avoid heat, these vegetables are normally planted either in the spring or in the fall. However, if you live in an area where the summers are cool (55°–70°F), you can plant these vegetables in the summer also.

Warm season vegetables usually produce a fruit of some kind. A lot of heat is needed to form the fruit. These vegetables are planted in early to mid-summer.

Some vegetables are normally planted as seeds outdoors. Others are normally planted as plants. The accompanying lists show which vegetables are in each group. Lettuce, melons, and onions are on both lists, as they are planted as seeds or plants.

Cool Season Vegetables

Asparagus	Chard
Beets	Lettuce
Broccoli	Onions
Brussels sprouts	Peas
Cabbage	Potatoes
Carrots	Radishes
Cauliflower	Spinach
Celery	Turnips

Warm Season Vegetables

Beans	Melons
Corn	Peppers
Cucumbers	Squash
Eggplant	Tomatoes

Vegetables Planted as Seed Outdoors

Beans	Melons
Beets	Onions (green)
Carrots	Peas
Chard	Radishes
Corn	Spinach
Cucumbers	Squash
Lettuce (leaf)	Turnips

Vegetables Planted as Plants

Asparagus	Eggplant
Broccoli	Lettuce (head)
Brussels sprouts	Melons
Cabbage	Onions (bulb)
Cauliflower	Potatoes
Celery	Peppers
	Tomatoes

When to Plant

Many home gardeners like to start seeds indoors for those vegetables which are normally transplanted outdoors as plants. These gardeners find a greater variety available, and have the added pleasure of "growing it themselves". If you are going to grow your own plants from seeds indoors, you will need to start the plants early so they will be ready on time.

There are, therefore, three different times we need to consider in determining when to plant:

1) the time to plant seeds outdoors, for those vegetables normally planted this way

2) the time to transplant the plants outdoors, for those vegetables normally planted this way

3) the time to start seeds indoors in order to have plants ready on time to transplant outdoors

These three times are shown in the table below for vegetables to which they apply. The time for no. 3 above is shown in the table as time to allow between planting seeds indoors and transplanting plants outdoors.

For many vegetables, you may want to transplant just after the spring frost. In this case, find the average date of spring frost in your area from the table, Page 4. Then count backward the required number of weeks shown in the last column.

For example, suppose the table, Page 4, shows your average last spring frost is April 1. For broccoli, you would count back 6 weeks to February 18, and plant your seeds indoors then.

Succession Planting

After the initial planting, succession plantings at 1–2 week intervals should be planned for certain vegetables. This spreads the harvesting time out accordingly. The vegetables for which succession plantings are commonly made are shown in the table below.

VEGETABLES PLANTED SUCCESSIVELY

Beans	Corn
Beets	Lettuce
Broccoli	Peas
Cabbage	Radishes
Carrots	Spinach
Celery	Turnips

When to Plant

PLANTING TIMES

Vegetable	Time to Plant Seeds Outdoors	Time to Transplant Outdoors	Time to Allow Between Planting Seeds Indoors and Transplanting Outdoors
Asparagus	N.A.	Early spring	N.A.
Beans			
Snap	Mid-spring to early summer	N.A.	N.A.
Lima	Early to mid-summer	N.A.	N.A.
Beets	Early to mid-spring and in the fall	N.A.	N.A.
Broccoli	N.A.	After spring frost and late summer to fall	6 weeks
Brussels sprouts	N.A.	After spring frost and late summer to fall	5 weeks
Cabbage	N.A.	After spring frost and late summer	6 weeks
Carrots	Early to mid-spring and in the fall	N.A.	N.A.
Cauliflower	N.A.	After spring frost and late summer to fall	6 weeks
Celery	N.A.	After spring frost and late summer	6 weeks
Chard	Early spring to late summer	N.A.	N.A.
Corn	Early to mid-summer	N.A.	N.A.
Cucumbers	Early to mid-summer	N.A.	N.A.
Eggplant	N.A.	Mid-summer to early summer	9 weeks

N.A. – Not Applicable

PLANTING TIMES (contd)

Vegetable	Time to Plant Seeds Outdoors	Time to Transplant Outdoors	Time to Allow Between Planting Seeds Indoors and Transplanting Outdoors
Lettuce			
Leaf	Early to late spring and late summer to fall	N.A.	N.A.
Head	N.A.	After spring frost and late summer	5 weeks
Melons	Mid to late spring	Early to mid-summer	9 weeks
Onions			
Green	Early to late spring and in the fall	N.A.	N.A.
Bulb	N.A.	After spring frost	8 weeks
Peas	After spring frost and in the fall	N.A.	N.A.
Peppers	N.A.	Mid-spring to early summer	8 weeks
Potatoes	N.A.	Early spring and late summer	N.A.
Radishes	Early to late spring and in the fall	N.A.	N.A.
Spinach	After spring frost and in late summer	N.A.	N.A.
Squash (both summer and winter)	Early to mid-summer	N.A.	N.A.
Tomatoes	N.A.	Mid-spring to early summer	8 weeks
Turnips	Early to late spring and late summer	N.A.	N.A.

N.A. – Not Applicable

▶ Planting Seeds in Open Ground

One or two days before planting, the seed bed should be watered thoroughly. The soil should become wet to a depth of ten inches or more.

When you are ready to plant, refer to your plan. It will tell you how much space you have allotted to each vegetable. It will also remind you of your plans for succession planting, if any.

Seeds are commonly planted in rows [1]. This makes a neat garden, and young plants are easier to distinguish from weeds. However, rows are not necessary. If you prefer, you can plant in wide bands [3] or scatter the seeds in broadcast fashion. Melon and squash seeds are usually placed in "hills" [2].

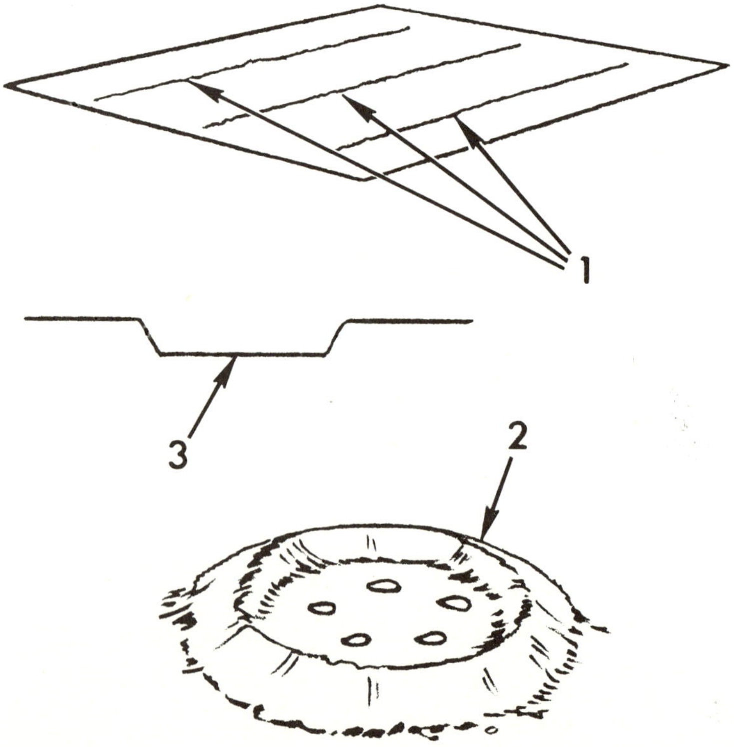

Planting Seeds in Open Ground

Seeds are of various sizes, ranging from very small (carrots, lettuce) to very large (beans, corn, peas). Depending mainly on size, seeds have a different depth of planting [3], distance between seeds in a row [1], and distance between rows [2].

These planting dimensions are usually given on the seed packet and should be followed. The accompanying table gives approximate values for these dimensions.

Vegetable	Depth to Plant Seed (inches)	Distance Between Seeds in Rows (inches)	Distance Between Rows (inches)
Beans (snap)	1	2	24
Beans (lima)	1	4	24
Beets	½	1	12
Carrots	½	½	12
Chard	½	1	18
Corn	1	3	24
Cucumbers	1	3	48*
Lettuce (leaf)	½	1	12
Melons	1	3	48*
Onions (green)	¼	½	12
Peas	1	2	24
Radishes	½	1	12
Spinach	½	1	12
Squash	1	3	48*
Turnips	¼	1	12

*Hills

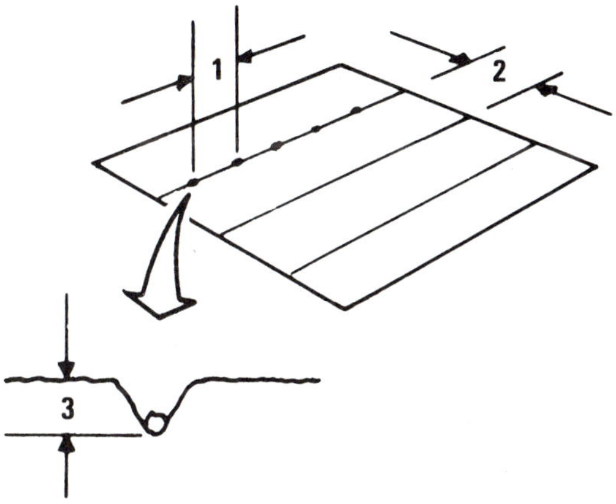

Planting Seeds in Open Ground

In planting seeds in a 4' x 4' module, the distances in the preceding table will determine how many seeds to plant. For example, lima beans [1] would be planted in two rows as shown. Later, these will be thinned (Page 31). The advantage of using a small module is again seen here: There is no need to walk on the soil which is to be planted. Thus, you avoid packing down the soil.

Cucumbers, melons, and squash are planted in the center of the module. A few seeds [3] are planted in each hill [2] about three inches apart.

An alternative method of planting cucumbers, melons, and squash is to grow them vertically on supports [5] (Page 34). If this is done, the seeds [4] are planted close to the bottom of the support, using the following approximate distances between seeds or seedlings:

cucumbers – 6 inches
melons – 24 inches
squash – 18 inches

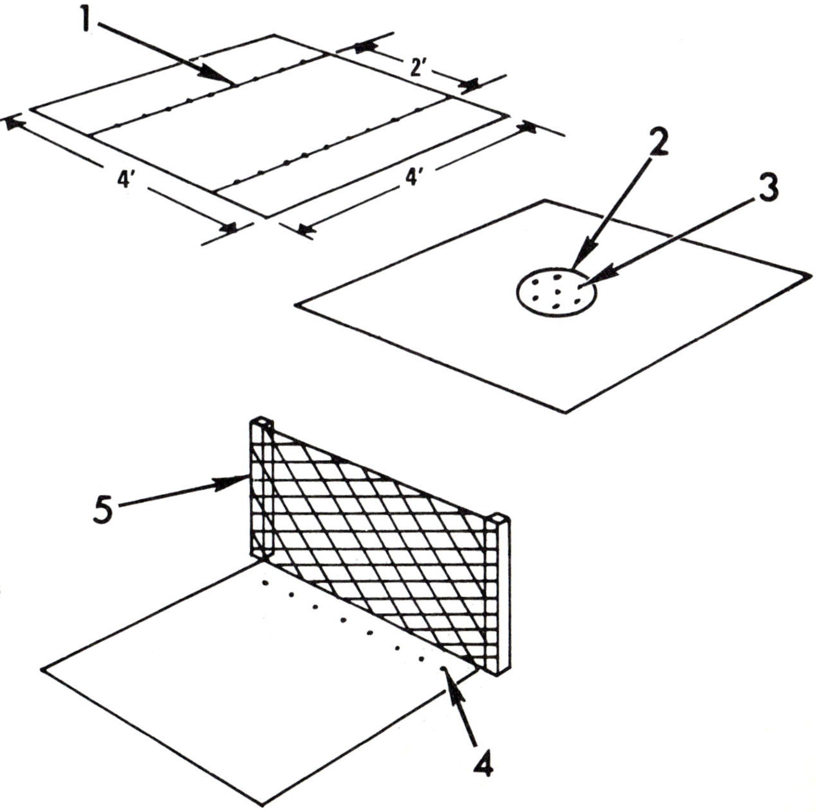

PLANTING

Planting Seeds in Open Ground

For smaller seeds, planted to a depth of 1/2 inch or less (see table, Page 29), a furrow can be made with the finger or a "V" indentation [2] can be made with the edge of a board [1].

For larger seeds, planted to a depth of 1 inch, a furrow can be made with the corner of a hoe [3]. A board [4] can be used as a guide to make a straight furrow.

The method of placing the seeds in a furrow also differs for small and large seeds.

Small seeds [7] can be tapped out of the seed packet [8], shaken from a salt shaker [5], or trickled out between thumb and forefinger. Small seeds are often mixed with sand or soil for ease of planting.

Large seeds [6] are placed in the furrow by hand one at a time.

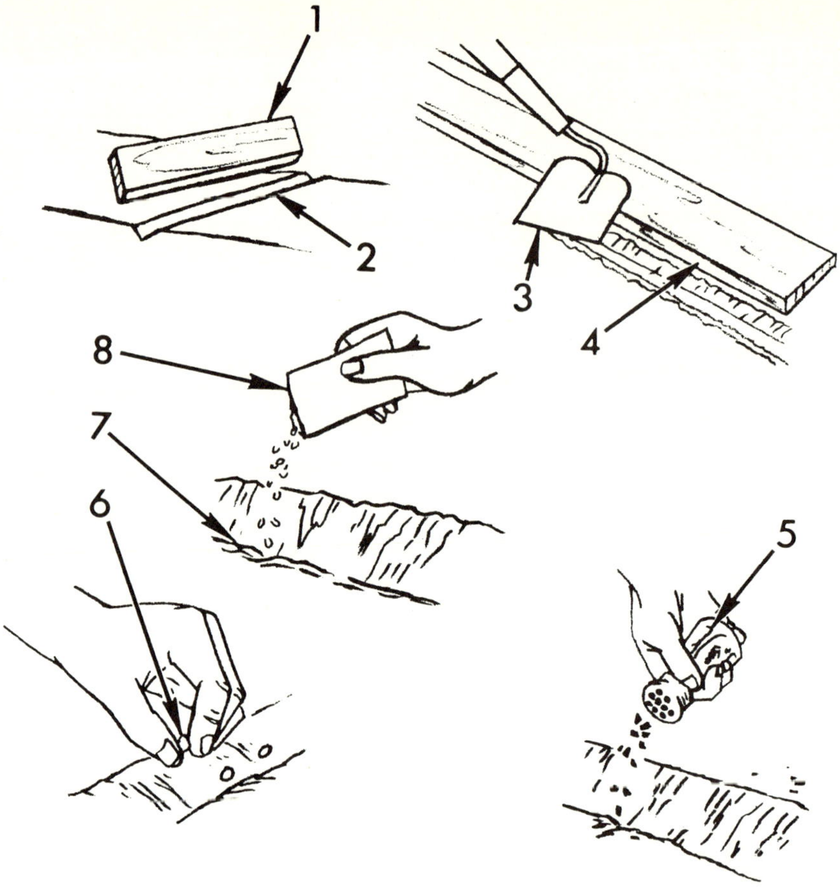

Planting Seeds in Open Ground

The procedure for planting seeds is as follows:

1. Using one of the methods suggested (section above), make furrow [1] of the depth and distance apart shown in the table (Page 29) or on the seed packet.
2. Using one of the methods suggested (section above), place the seeds [2] in the furrows.
3. Using a hoe or your hands, push the soil back into the furrows [5] covering the seeds.
4. Using a hoe or your hands, press the soil [3] down gently to make firm.
5. Using a watering can or a hose with a fine spray [4], water thoroughly to a depth of 6 inches.

The soil should be kept moist while the seeds are sprouting. Use a sprinkler or fine mist spray to water daily.

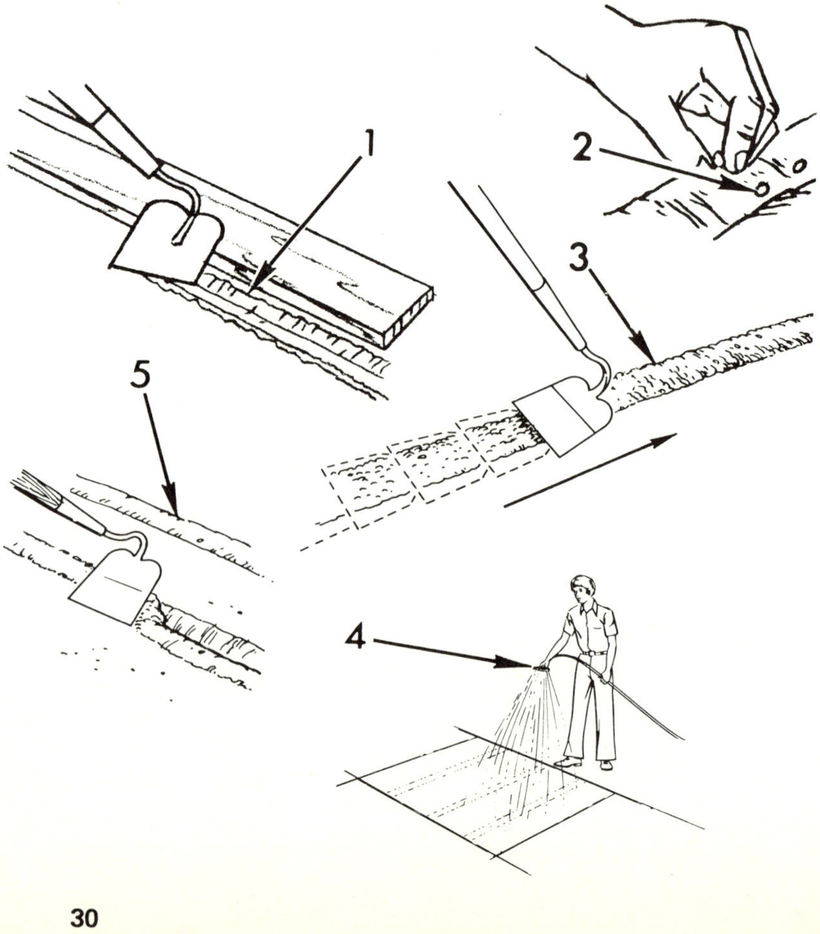

Planting Seeds in Open Ground

Hill planting: As mentioned earlier, large vegetables such as cucumbers, melons and squash are often planted in "hills". The seeds are planted in a circle [1], 12–18 inches across. After covering the seeds, a small embankment [2] is formed to form a basin for holding water.

Marker: After planting, the seed packet [3] or other convenient marker can be used to mark the rows [4]. Otherwise you may not remember what is planted. This is necessary, for example, in proper thinning (see below).

Thinning: Most gardeners purposely plant seeds somewhat too thickly, for insurance. When the seeds sprout, they are too close together and must be thinned to allow proper development.

The time to thin is when the seedlings [6] are about three inches high and have two pairs of leaves [5]. Seedlings which are pulled up can be used in salads.

The space to allow between plants after thinning is shown in the table.

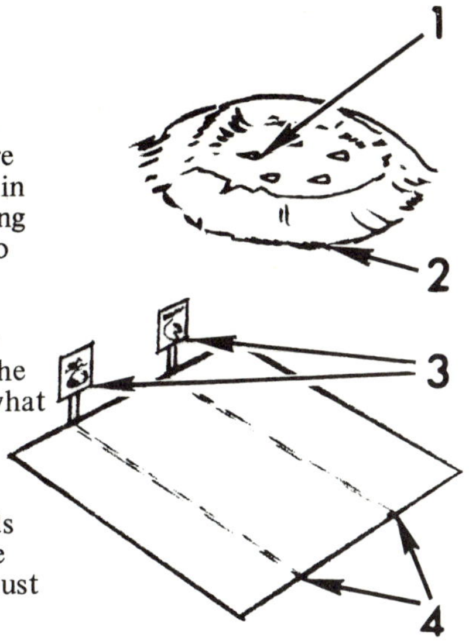

Vegetable	Space Between Plants After Thinning (Inches)
Beans (snap)	6
Beans (lima)	12
Beets	3
Carrots	2
Chard	12
Corn	12
Cucumbers	48*
Lettuce (leaf)	6
Melons	48*
Onions (green)	1
Peas	3
Radishes	1
Spinach	8
Squash	48*
Turnips	6

*Hills 48" apart with 2–3 plants per hill

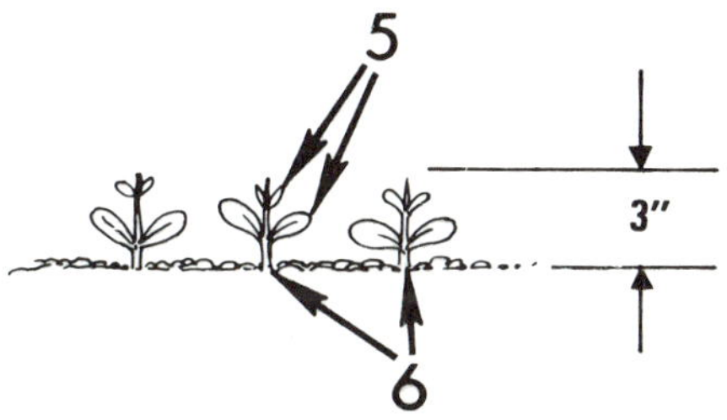

▶ Starting Seeds Indoors

Seeds can be started indoors in containers if you prefer growing your own rather than buying transplants. Starting seeds indoors can also be a help if you have a short growing season, because it gives you extra growing time. Also, you can save money and choose from a greater variety of vegetables.

The Planting Mix

The planting mix should be a commercial sterilized potting soil. Soil from the garden should not be used unless it is first sterilized, to prevent "damping off" disease. You can sterilize your own soil by baking it in a shallow pan in the oven for an hour at about 175°F.

A widely used substance for starting seeds is vermiculite. It is a relatively inexpensive, inert material with excellent properties. Also, there are many potting soils available on the market which are suitable.

The Container

Many types of containers can be used for starting seed indoors. A common type which works well is a flat, shallow aluminum foil pan [2]. Holes [3] should be poked in the bottom to allow drainage. Other commonly-used containers include wooden

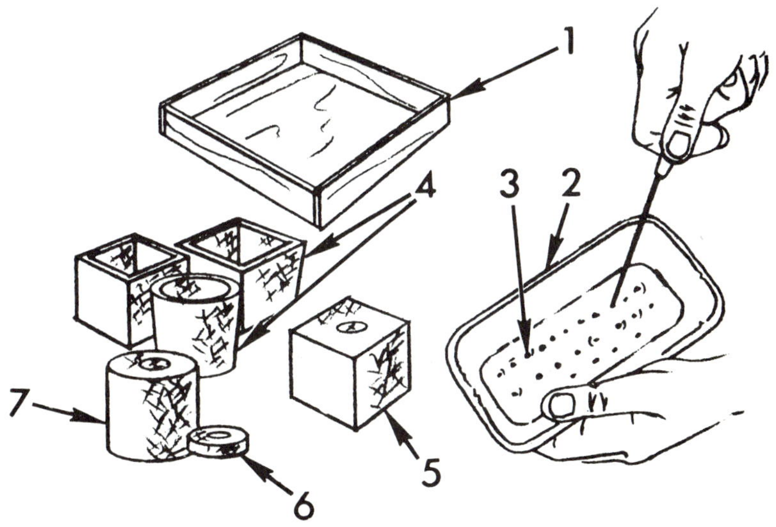

flats [1], cottage cheese cartons, milk cartons, ice cream cartons, etc. Except for wooden flats, drainage holes should be poked with a fork or ice pick.

A popular type of container is made from pressed peat. These are available in different sizes and shapes: pots [4]; discs [6], which swell to cylinders [7] when wet; and cubes [5]. An advantage of these containers is that the shock of transplanting is minimized, since the container stays with the plant until it decomposes.

PLANTING

Starting Seeds Indoors

The procedure for starting seeds in a shallow pan, flat, or similar container is as follows:

1. Fill container [1] with planting mix [2] to within about one-half inch of top.

2. Lightly sprinkle seed [3] on top of planting mix. Cover seeds with shallow layer (1/8 to 1/4 inch) of planting mix.

Watering may be done either with a watering can or by immersing container in a tray of water.

3. Watering

   a. Watering can – Water planting mix carefully until saturated.

   b. Immersing – Place container [1] in tray of water. Allow water to soak up through drainage holes until planting mix is saturated. Remove container [1] from water and allow to drain for a few minutes.

A plastic bag may be used to help germination. Using a plastic bag eliminates the need for additional watering. If using plastic bag, go to Step 4. Otherwise, go to Step 5.

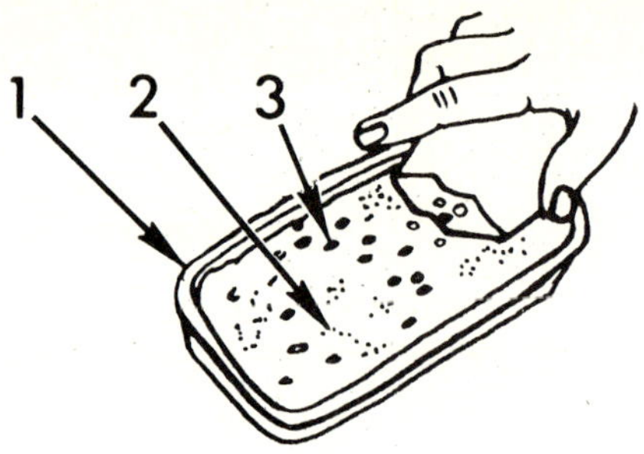

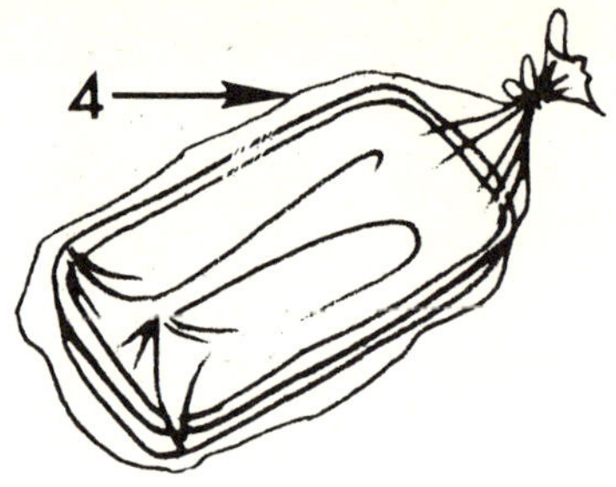

4. Using plastic bag

   a. Place container [1] in plastic bag [4] and tie bag shut.

   b. Place container in the bag in warm (above 60°F) spot out of strong light.

   c. When seedlings appear, remove container from plastic bag and place in sunny area. Go to Step 6, below.

5. Not using plastic bag

   a. Place container [1] in warm (above 60°F) spot out of strong light.

   b. Water planting mix daily to keep moist.

   c. When seedlings appear, place container in sunny area. Go to Step 6, below.

Starting Seeds Indoors

6. Check moisture daily. Water sufficiently to keep planting mix moist but not soggy.

To provide more room, seedlings must be transferred to new container (5,6) when second pair of leaves [2] appears.

A kitchen knife or fork is useful for taking seedlings out of old container [1].

CAUTION

Do not hold seedling by stem [3] —hold by leaf [4]. Do not shake off soil on roots.

7. When second pair of leaves [2] appears, take seedling gently out of pan with knife or fork or with fingers.

8. Using finger, make hole in soil of separate container [6] or larger container [5]. Transplant seedling to container.

Repeat Steps 7 and 8 for each seedling.

About 10 days before plants are to be placed outdoors, start the "hardening" process as described in the next three steps.

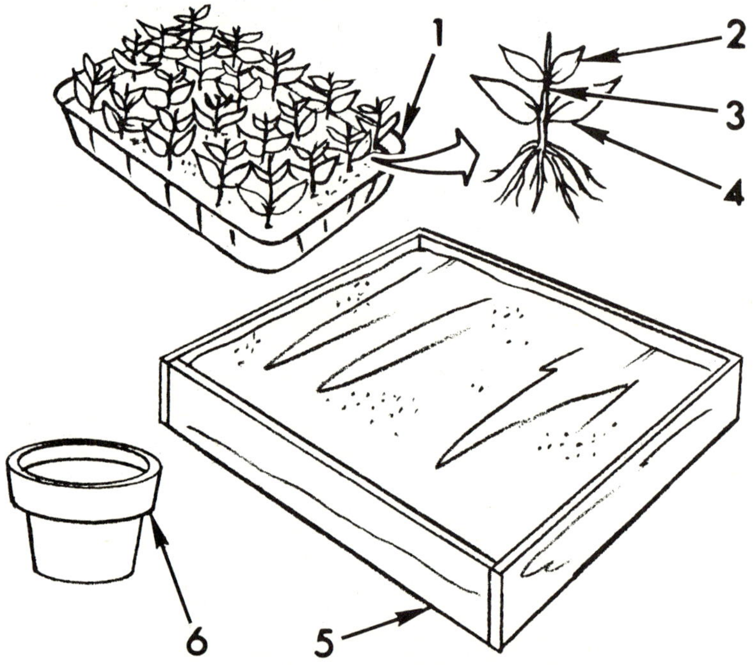

9. Put out seedlings during day in partial shade, bring in before sundown.

10. Gradually increase exposure to sunlight.

11. Leave out all night before planting outdoors.

Starting Seeds Indoors

Peat pots: Peat pots can also be used for growing seeds indoors. Peat pots are small individual pots [1], discs [2], or cubes [4]. They are made of compressed peat moss and can be bought at nurseries, hardware stores, etc.

Some types are already filled with planting mix. Others you must fill yourself. The disc-type [2] requires no planting mix. It expands when wet into a cylinder [3]. The cube [4] also needs no planting mix.

There are several advantages of using peat pots, over the transplant procedure just described (Page 32).

1) Less work is required, since the seedlings do not need to be transferred to a larger container.

2) There is no transplant shock, since there is no transfer to a larger container and since the seedling is eventually planted, pot and all, in the garden (see Transplanting Plants, below).

3) Nutrients are usually built right into the pot.

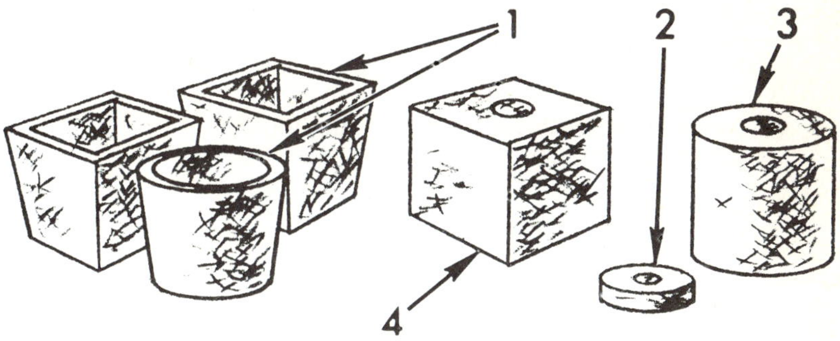

If peat pots are used, follow these procedures:

1. Add planting mix to pot [1], if necessary.
2. Place a few seeds in each pot.

A plastic bag can be used to keep the pot moist, as before (Page 32).

3. Keep each pot moist and out of strong light.
4. When seedling appears, place in sunny area and water daily.

Now use the "hardening" process described earlier (Steps 9–11, Page 32).

▶ Transplanting Plants

Seedlings grown indoors should be "hardened" (Page 32), before transplanting outdoors. Plants which are bought will normally be ready to transplant outdoors without requiring the hardening process.

For transplanting plants in the garden, proceed as follows:

Tools:

Trowel [5]

1. Water plants thoroughly while still in container [1].

The hole [4] in which the plant is to be transplanted should be large enough and deep enough that plant and peat pot (if any) will fit in easily. Roots should be disturbed as little as possible.

2. Using trowel [5], dig hole [4] in garden.

Most plants should be set in to the same depth or slightly deeper than they were in the container. Tomato plants should have about half the stem [2] under ground. If plant is in peat pot, do not remove from pot.

3. Carefully place plant in hole.

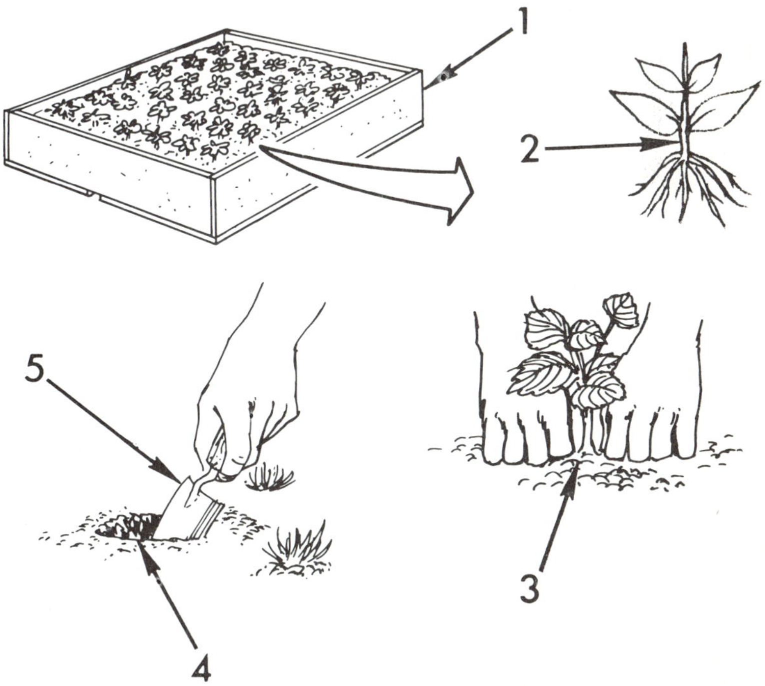

4. Cover roots with soil [3] and press gently, but firmly.
5. Water the plant thoroughly, to a depth of at least 6 inches, to settle soil around roots.

PLANTING

Transplanting Plants

The distance to provide between plants in a row, and the distance between rows, are shown in the table.

In planting plants [1] in a 4' x 4' module, for example, broccoli, brussels sprouts, cauliflower, eggplant, and tomatoes should be planted in four places as shown. A melon plant [2] should be placed in the center of the module. If other melon plants are planted, they should each occupy a separate module.

Vegetable	Distance Between Plants in Row (inches)	Distance Between Rows (inches)
Asparagus	18	24
Broccoli	24	24
Brussels sprouts	24	24
Cabbage	12	24
Cauliflower	24	24
Celery	6	18
Eggplant	24	24
Lettuce (head)	12	18
Melons	48	48
Onions (bulb)	4	12
Peppers	18	24
Potatoes	12	24
Tomatoes	24	24

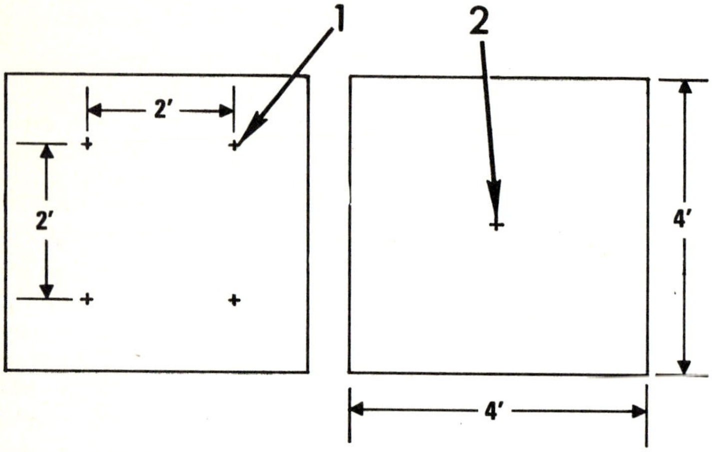

▶ Providing Structural Support

A means of support is usually provided for pole beans, cucumbers, peas, and tomatoes. Often, to conserve space, support is also provided for melons and squash.

Beans and peas are light and can be supported by string. Examples of support for these vegetables are shown [1,5,6].

Cucumbers, tomatoes, melons, and squash require sturdier support. These vegetables can be supported on posts with attached boards [4]. As the vines grow, tie them to the boards. Melons should be supported with cloth slings [3].

Cucumbers, tomatoes, melons, and squash are also commonly supported on lath trellises or chicken wire fences [2]. A 4' x 4' post is placed at each end, and the trellis or chicken wire is supported by the posts.

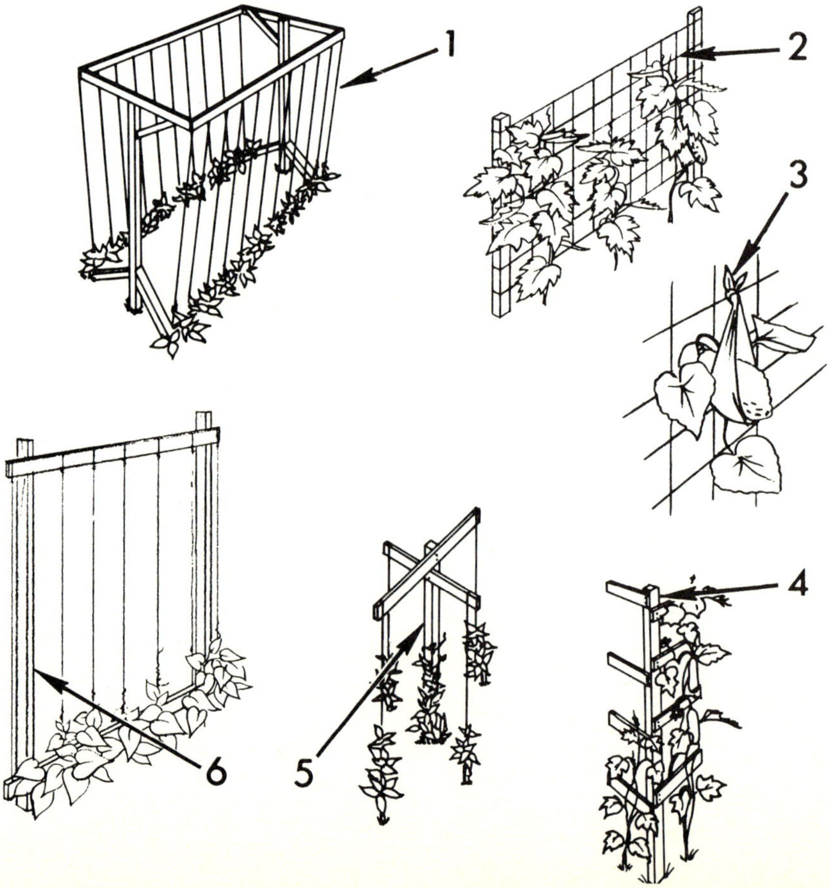

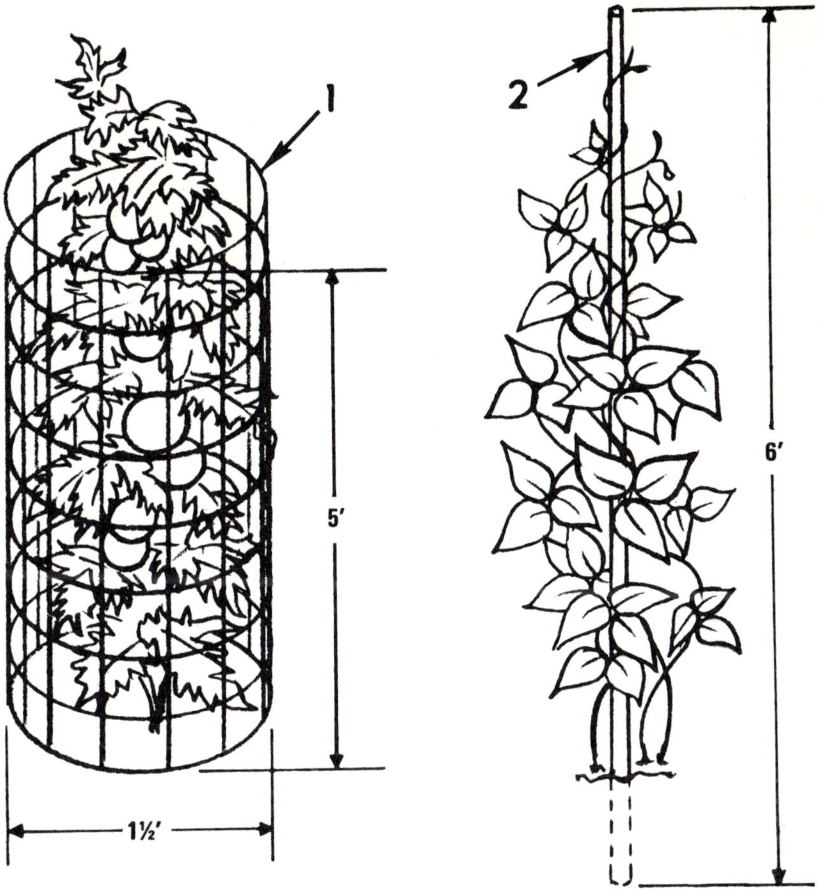

Providing Structural Support

Another effective support for the heavier vegetables is a ring of construction wire [1]. The ring should be about 5 feet high. A 5-foot length of construction wire, when formed in a circle, will produce a ring about 1-1/2 feet in diameter.

Tomatoes are often supported by a simple stake [2]. The stake is a 1" x 1" or 2 " x 2", about 6 feet long. Such stakes are also used for beans and peas.

GARDEN CARE

▶ Watering

Proper watering is important to a good vegetable crop. First, water is needed as a basic nutrient for the vegetables to grow. Second, water is the means by which other nutrients are made available to the plants.

It is difficult to give precise rules for watering. The general rule is to water deeply and regularly, but not too frequently. Roots must "breathe", and they cannot do so if they are continuously surrounded by water.

The ideal is to water in such a way that the water reaches a depth of 2 to 3 feet, particularly where the larger vegetables are planted, and then let the soil become almost dry to a depth of about 6 to 8 inches before watering again.

The way to water properly will depend to some extent on the watering method used, the climate, the kind of soil, and whether mulch is used (Page 37).

Typically, watering should be performed weekly for one to two hours. Depending on your particular circumstances, however, you might find it necessary to water twice a week or only once every two weeks.

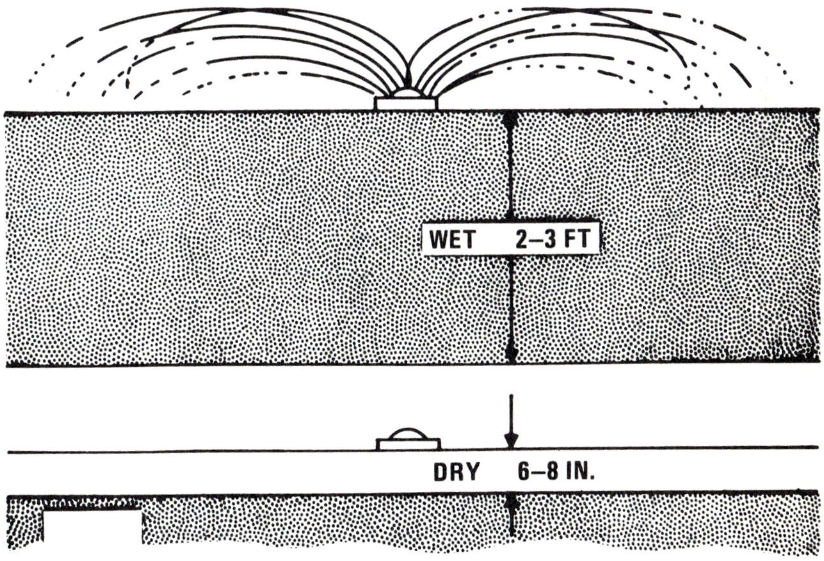

The watering guidelines above apply to established plants. When seeds are first planted, daily watering will be necessary to keep the soil moist until the seeds germinate (see Page 30). Also, small seedlings will require frequent watering. Deep soaking is not needed since their roots are quite shallow. As the seedlings grow in size, water less frequently and more deeply until you are watering about once a week to a 2–3 foot depth.

GARDEN CARE

Watering

Watering methods include a hose [1] on the ground, a sprinkler [5], and a soaker [3]. The water force of a hose on the ground can be reduced by tying a piece of burlap [2] over the end of the hose, or a "bubbler" [4] can be bought from a garden store.

If a sprinkler is used, watering is best done in the morning so that leaves can dry during the day. Otherwise, there is the possibility of mildew or rust. Also, watering with a sprinkler can damage hot weather vegetables such as squash or tomatoes. It may cause the fruit to crack. Therefore, after the fruit appears, it is best to change to a hose on the ground.

After watering, push a stake [7] or rod into the soil. It should go in easily 2 to 3 feet. If not, continue watering. If you have hardpan, rock, etc., less than 2 feet down, water until the soil is thoroughly wet down to the hard surface.

Three to four days after watering, use a trowel [6] to check soil. If soil is almost dry to a depth of about 6 to 8 inches, it is time to water again. If not, check every day or two until the soil is nearly dry to about this depth. Then water again thoroughly.

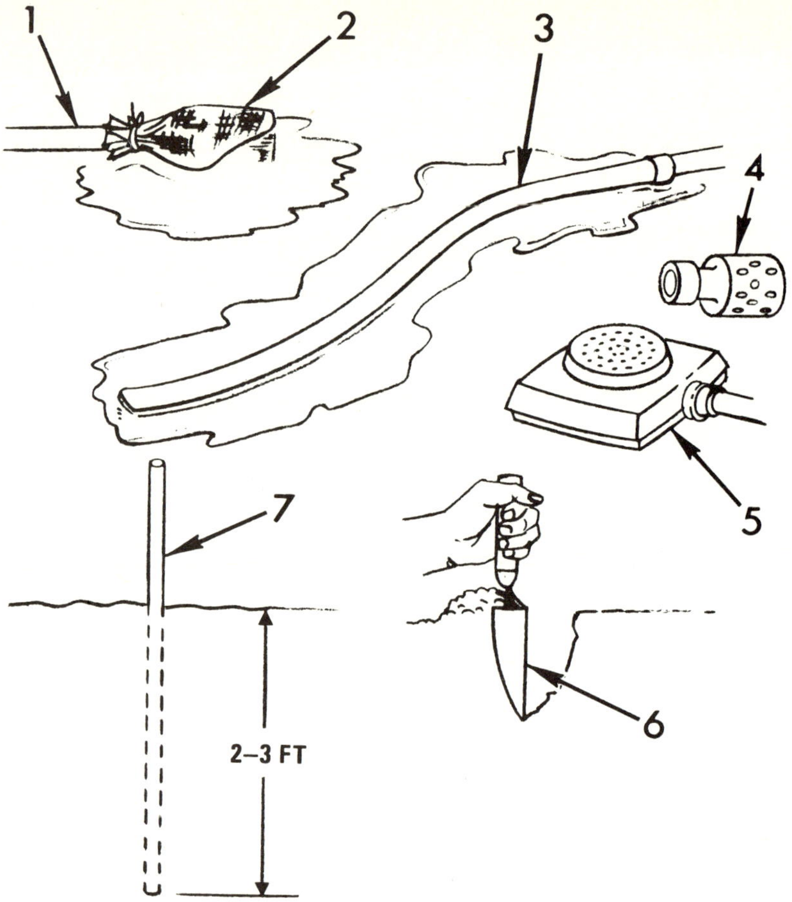

▶ Fertilizing and Mulching

Fertilizing

If the soil is prepared properly as described in Preparing the Soil (Page 21), there will not be much need for adding fertilizer as the vegetables grow. The method described incorporates a considerable amount of fertilizer into the soil. Also, if a timed-release fertilizer was used, it should last during the growing season.

However, some vegetables require more fertilizer than others. These vegetables can be helped in their growth by an application of fertilizer during the growing period, if a timed-release fertilizer was not used.

The vegetables which can use extra fertilizer during their growth are:

Beans	Eggplant
Broccoli	Melons
Brussels Sprouts	Onions
Cabbage	Peppers
Celery	Spinach
Corn	Squash
	Tomatoes

About a month to six weeks after the plants appear or are planted, a fertilizer should be applied to the above vegetables.

A fertilizer should be used which supplies nutrients fairly quickly to the plants. A good organic fertilizer for this purpose is fish emulsion. Or, a standard commercial fertilizer such as 5-10-5 may be used.

Whatever fertilizer is used should be applied according to instructions attached to the fertilizer. The fish emulsion comes in bottles. It is diluted and applied with a watering can [1]. Commercial fertilizers are in granular form. They are either broadcast [3] near the plants or placed in shallow furrows [2] which are then covered with soil. If broadcast, they should be worked gently into the top of the soil with a rake.

Fertilizing and Mulching

Mulching

"Mulching" refers to any number of materials which are spread on the top of the soil. Such materials include common materials, usually organic, such as hay, straw, sawdust, woodchips, even newspapers.

There are many advantages to using a mulch. Mulches reduce or eliminate the growth of weeds. Watering can be less frequent, as there is less evaporation from the soil surface. Vegetables will be cleaner, particularly those which bear fruit such as squash and tomatoes. The danger of frost damage is reduced as the mulch tends to keep the soil warmer. Finally, most mulches (including newspapers) will decompose in time and add humus and nutrients to the soil.

The mulch [2] is spread around the vegetable plants, away from the stems [1]. The mulch should be 2 to 3 inches thick. Good mulches to use are hay, straw, sawdust, or woodchips. Leaves should not be used unless first put through a shredder. Grass clippings, if used, should be spread thinly – about one-half inch.

Weeds

As mentioned above, one advantage of mulching is that weed growth is reduced. Whether or not mulch is used, all weeds should be pulled. Weeds compete with vegetable plants for moisture and nutrients.

Pull weeds regularly, while they are still small. Otherwise, you will disturb the vegetable plants or even dislodge them by mistake.

▶ **Controlling Insects and Diseases**

Controlling Insects

Insects are best controlled through simple prevention, rather than "find and destroy". If the vegetables are given proper care, they will grow fast and there should be little trouble from insects.

Cleanliness is important in prevention. Avoid leaving piles of lumber, garden refuse or other debris near the garden. Such piles serve as breeding and hiding places for insects. No vegetables, leaves, etc., should be left on the ground where they fall – they should be cleaned up. Such simple measures can help prevent insects.

Another preventive found effective is to plant marigolds [1] and other aromatic plants, such as onions, garlic, or hot peppers, around the garden. The strong odors seem to repel insects. The roots of marigolds secrete a substance that kills nematodes, a type of parasitic worm that causes lumps [2] on the roots of plants.

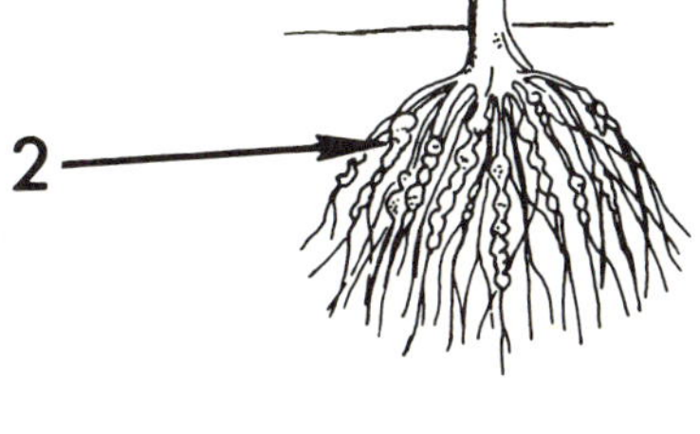

GARDEN CARE

Controlling Insects and Diseases

If prevention fails and you discover insects on your vegetables, it is best to proceed in stages. Do not immediately spray with an insecticide—this should be considered a last resort. When used, insecticides should be for a specific vegetable and specific insect, as described later (see table, Pages 39—40).

As a first step, try to pick off the insects by hand or spray them off with a hose. This works well for larger, non-winged insects such as worms, caterpillars and beetles.

Second, try using one of the following simple sprays:

1) For aphids, use a soap and water spray. Mix 3 tablespoons of soap flakes (not detergent) per gallon of water. Use a spray can [1] or tank sprayer [2].

2) For other insects, use a spray made from plants. Using a blender, mix strong-smelling plants such as onions, garlic, hot peppers or marigolds with an equal amount of water. Strain off the solids and discard. Use 3 tablespoons of liquid per gallon of water in a sprayer.

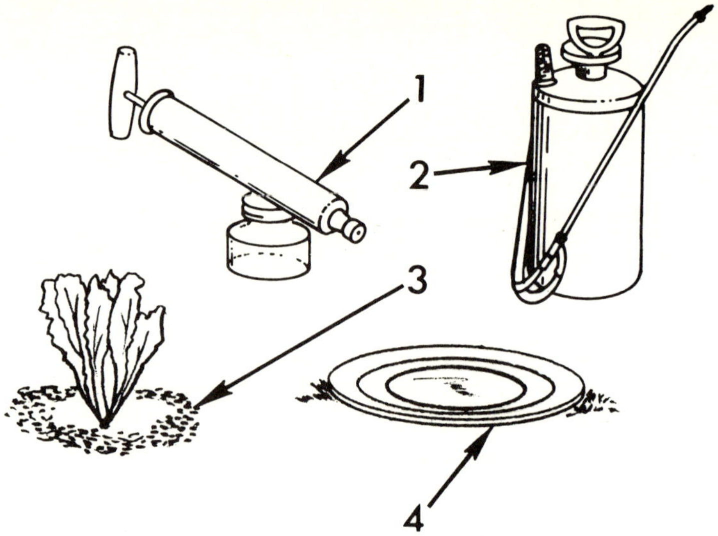

Third, special measures can be used against specific insects.

Snails and slugs eat the leaves, and leave a silver trail behind.

For snails and slugs, place saucers of beer [4] outside at night. The beer attracts the snails and slugs, which drown. As an additional precaution against snails and slugs, ashes [3] can be placed around plants.

Controlling Insects and Diseases

The presence of cutworms is indicated when small plants are cut off at ground level [1]. Root maggots are indicated if plants droop and there are holes in the parts of the plant that are underground.

For cutworms, place a stiff paper collar [2] or a tar paper collar around the base of the plant. If a paper collar is used, it should extend into the soil. The tar paper collar [3] lies flat on top of the ground. The tar paper collar can also be used against root maggots.

The corn earworm is common around corn, and is indicated by missing kernels [4] or silks [5] in the ears. To reduce damage, the silks can be cut off after the ears reach full size.

The presence of squash vine borer is indicated by holes in the stem and wilted runners of squash plants. The hole in the stem is located by excrement left by the borer. Slit the stem with a knife or razor, remove the borer and heap the earth back up over the slit.

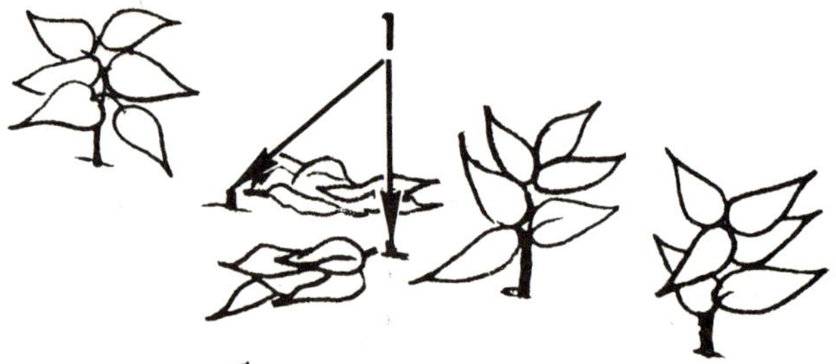

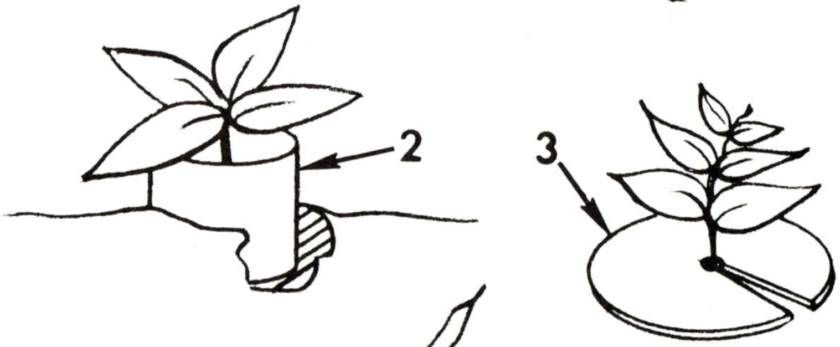

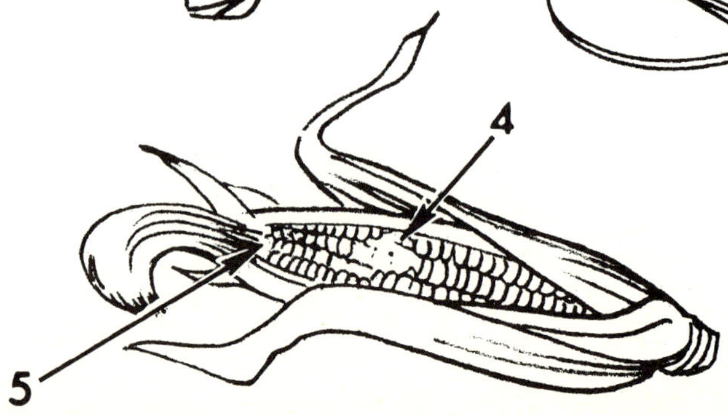

Controlling Insects and Diseases

Insecticides: If the procedures described fail to eliminate harmful insects, insecticide sprays may be purchased. These must be used very carefully, according to the written instructions provided. Some of those described below are also available as powders for dusting.

The least harmful to humans are the botanical sprays: rotenone, pyrethrum, and ryania. These sprays are derived from plants.

Chemical insecticides must be used with particular care, and include diazinon, malathion, and sevin.

A biological insecticide useful for worms and caterpillars is thuricide.

Helpful insects: Certain insects are useful for vegetable gardeners because they prey on harmful insects. These helpful insects, which can be purchased, include ladybugs [1], lacewing flies [2], praying mantises [3], and others.

Summary table: The table below and on Page 40 summarizes the types of harmful insects, the vegetables which they attack, signs,or indications of each insect, and various possible remedies.

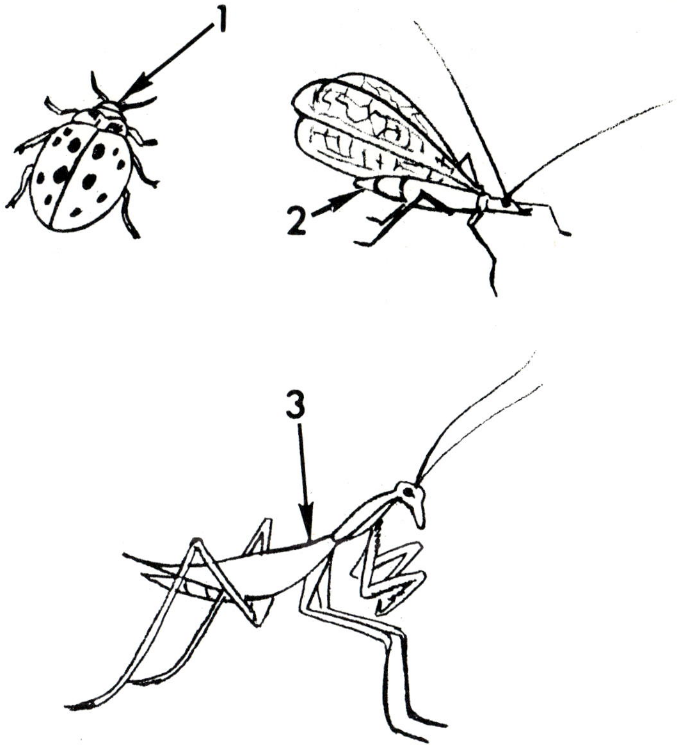

Controlling Insects and Diseases

COMMON INSECTS AND REMEDIES

INSECT	VEGETABLES	SIGN	REMEDY: Pick Off	Pyrethrum	Rotenone	Ryania	Diazinon	Malathion	Sevin	Thuricide	REMARKS
Aphid	Beans, broccoli, brussels sprouts, cabbage, cauliflower, chard, cucumbers, lettuce, melons, peppers, spinach, squash, tomatoes, turnips	Clusters of small insects on leaves		X	X	X		X			Spray with water or soapy solution
Asparagus Beetle	Asparagus	Leaves eaten	X	X	X	X			X		
Bean Leaf Beetle	Beans	Circular holes in leaves	X	X	X	X	X	X	X		
Cabbage Worm	Broccoli, brussels sprouts, cabbage, cauliflower	Holes in leaves	X		X					X	
Celery Worm	Celery	Leaves eaten	X							X	
Colorado Potato Beetle	Eggplant, peppers, potatoes, tomatoes	Leaves eaten	X	X	X	X	X	X	X		
Corn Borer	Corn	Holes in ears and stalks	X								
Corn Earworm	Corn	Silks and kernels missing	X			X					Cut off silks early or use mineral oil on silks
Cucumber Beetle	Cucumbers, melons, squash	Leaves and vines eaten	X	X	X	X	X	X	X		
Cutworm	Beans, broccoli, brussels sprouts, cabbage, corn, peppers, tomatoes	Small plants cut off					X		X		Use stiff paper collar or tar paper collar
Leaf Hopper	Beans, lettuce, potatoes	Wedge-shaped insects on leaves		X	X	X		X	X		
Leaf Miner	Spinach	Holes in leaves			X						Remove infested leaves
Mexican Beetle	Beans	Lower surface of leaves eaten	X	X	X	X	X	X	X		

Controlling Insects and Diseases

COMMON INSECTS AND REMEDIES (CONTD)

INSECT	VEGETABLES	SIGN	REMEDY: Pick Off	Pyrethrum	Rotenone	Ryania	Diazinon	Malathion	Sevin	Thuricide	REMARKS
Onion Maggot	Onions	Plants droop; holes in underground part of plants						X			Put ashes around base of plant
Onion Thrip	Onions	Small yellow insects at base of leaves		X	X	X		X			
Pea Weevil	Peas	Blooms eaten, holes in pod	X				X				
Root Maggot	Broccoli, brussels sprouts, cabbage, cauliflower, radishes, turnips	Plants droop, holes in underground part of plant					X				Use tarpaper collar
Snails & Slugs	Beets, lettuce	Leaves eaten, silver trail	X								Put ashes around plants; set out saucers of beer at night
Squash Bug	Melons, squash	Plants wilted		X	X	X					
Squash Vine Borer	Squash	Holes in stem, runners wilted									Locate hole in stem from excrement, slit stem with knife or razor, remove borer, heap earth back up over slit
Tomato Hornworm	Tomatoes	Leaves eaten	X								
Whitefly	Beans, tomatoes, eggplant	Scaly insects on underside of leaves; white, fluttering adults			X			X			

Controlling Insects and Diseases

Controlling Diseases

As with insects, the best protection against disease is prevention. Again, prevention starts with cleanliness. The garden should be kept free of piles of refuse, boards, fallen vegetables, and other debris.

Many seeds and transplants are disease resistant. Using these varieties will greatly decrease the danger of disease.

Cabbage, eggplant, and tomatoes should not be planted in the same locations in successive years. Changing the locations of these vegetables will reduce the danger of soil-borne fungus diseases.

The most common diseases are:

Mildew — gray powdery coating on leaves and stems

Rust — white or yellow warts on underside of leaves, then powdery red spores carried by wind

Blight and scab — spreading red, brown, or yellow spots, usually on lower leaves

A disease, if it occurs, will not usually affect every vegetable. It will be specific to a particular vegetable, and usually only a few plants will be affected. These plants should be destroyed, preferably by burning. Do not place them in the compost pile, if you have one.

▶ Harvesting

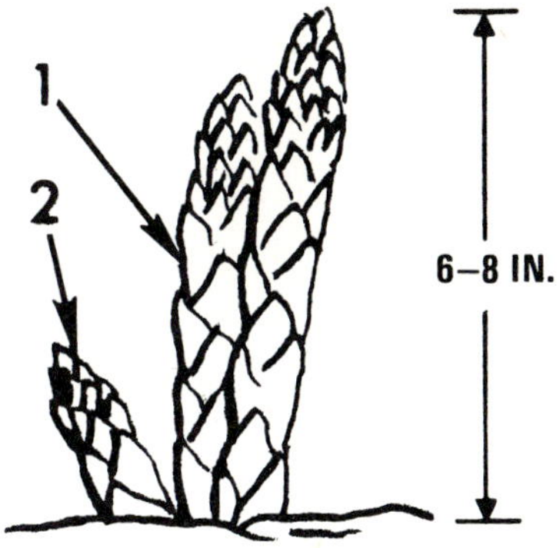

Vegetables will be considered individually in terms of when and how to harvest.

Asparagus: This is the only perennial on the list. There will be no harvest the first year plants are planted, a few spears the second year, and a full harvest the third year. Cut or snap off the spears [1] at ground level when they are 6 to 8 inches long. The spears should be tight at the tip [2], with no spreading. The quality of asparagus, like corn, deteriorates rapidly after harvest. Therefore, cook as soon as possible after cutting.

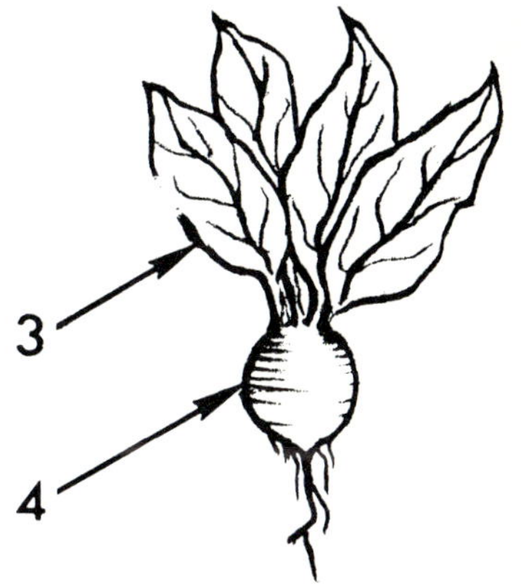

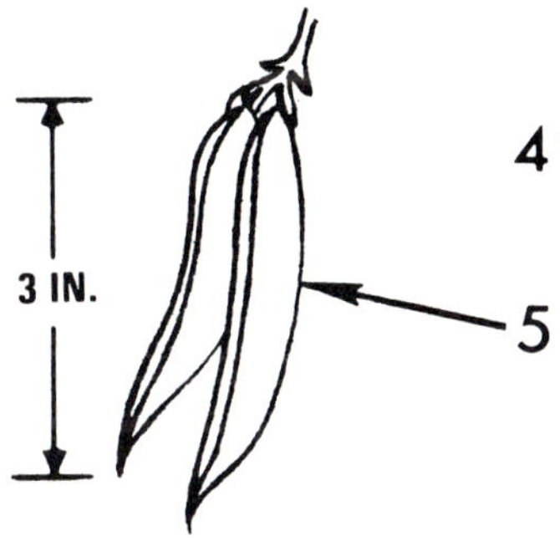

Beans: Beans are ready to pick when the pods [5] are about 3 inches long, are well rounded, and snap easily if bent. All pods should be picked when ripe, as this increases pod production by the plant.

Beets: Beets should be dug up when the root [4] is about 1-1/2 to 3 inches in diameter. Small beets are tender and the tops [3] can also be eaten. Larger beets tend to be tough and the tops less tasty.

Harvesting

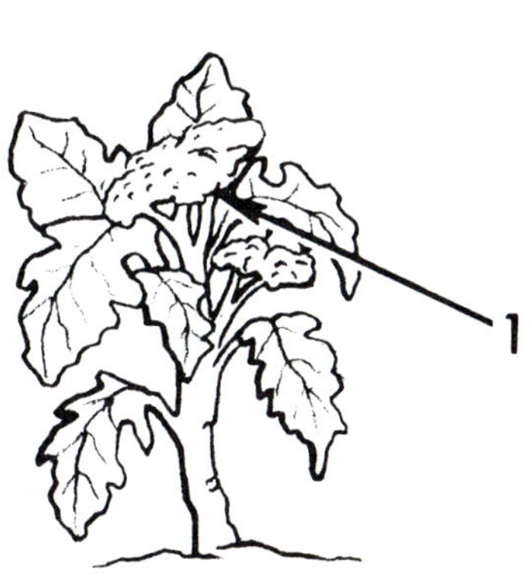

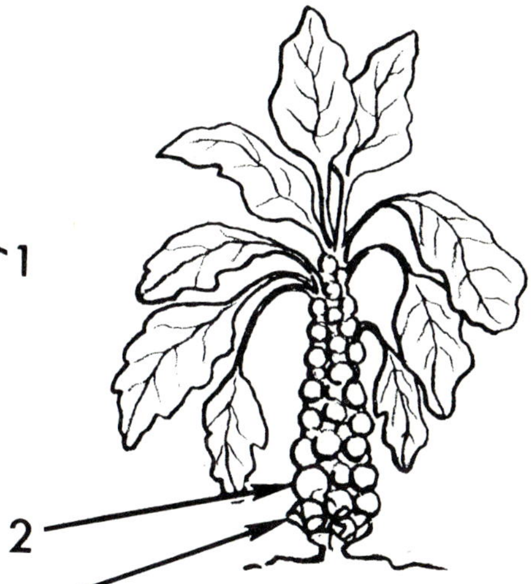

Broccoli: Cut below the head [1] before the head cracks open. Up to 6 inches of stem can be included and eaten. There will be continual harvest from side shoots.

Brussels Sprouts: Sprouts should be picked when they are still green and 1 to 1-1/2 inches in diameter. They can be twisted, snapped, or trimmed off. The lowest sprouts [2] should be picked each time, and the leaves [3] below the picked sprout should be broken off.

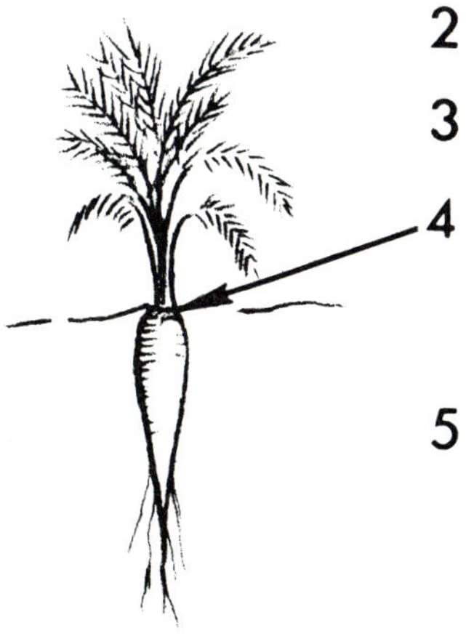

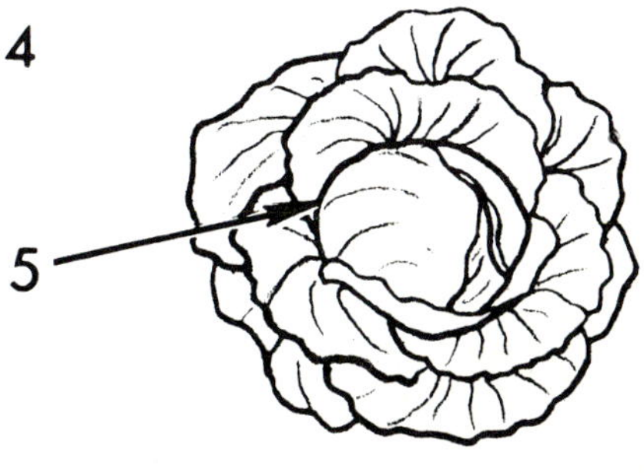

Cabbage: Pick as soon as the head [5] feels solid. Cut just below the head, leaving a few leaves to support new heads later.

Carrots: A small orange crown [4] appears when carrots are ready to harvest. If carrots are dug up while small, they are less woody than larger carrots. Other than this, they can be left in the ground and harvested as needed.

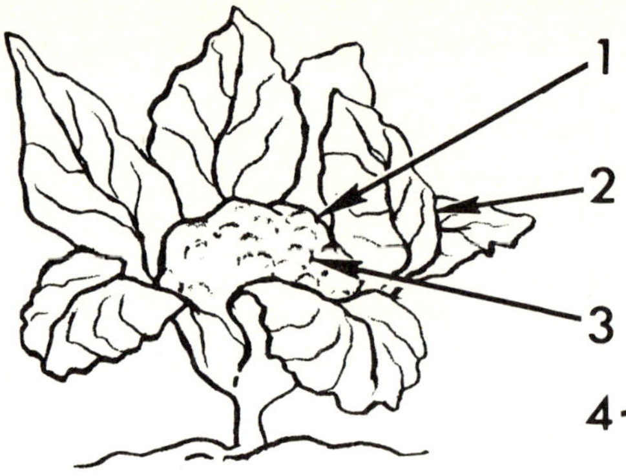

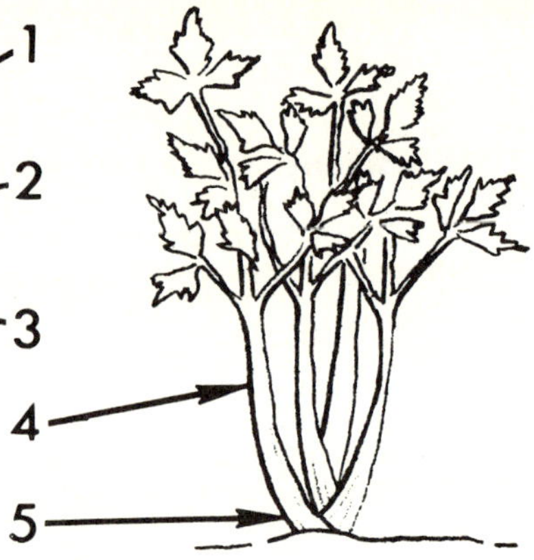

Harvesting

Cauliflower: When the head [1] is about the size of an apple, loosely tie the outer leaves [2] together with string. This will blanch the head, making it white. Do not tie the leaves too tight, as air should circulate. Cut after head is white, in one or two weeks, and before curds [3] begin to separate.

Celery: Cut plants at base [5]. Remove some of the outer leaves [4] as needed. Or, cut off entire plant just above roots. Smaller new stalks will grow from roots.

Chard: Cut off outer leaves [7] about two inches above ground, every few days. Leave a few center leaves to help the plant in continued growth.

Corn: Use a fingernail to pierce a kernel [6]. If milky juice spurts out, the corn is just right for picking. If the juice is clear, the corn is not yet ready. If the juice is pasty, the corn is past its prime. Plan the picking time so you can cook the corn immediately after picking. Corn starts losing flavor quickly after picking.

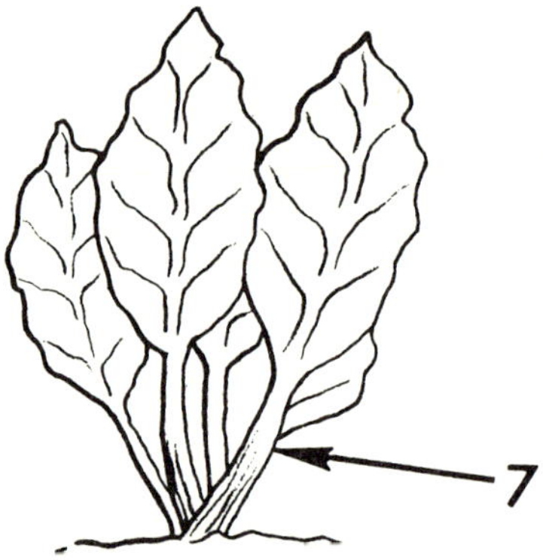

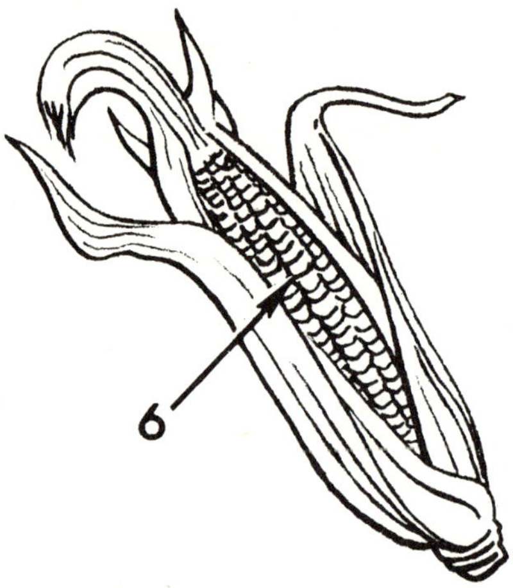

Harvesting

Cucumbers: For slicing, pick cucumber [1] when 6 to 8 inches long and while still green. As it turns yellow, the seeds become hard. Continue picking ripe cucumbers. This will help the growth of new fruit. If planning on pickling, pick fruit earlier, when 2 to 5 inches long.

Eggplant: Pick while there is a glossy shine on the surface [3]. Keep picking to help growth of more fruit. Since the stem [2] is woody, cut, do not pull.

Lettuce: Head lettuce [5] is harvested when the center feels firm. The entire plant can be pulled up. Leaf lettuce [4] is harvested a few leaves at a time, or if the plant is cut off 1 to 2 inches above ground, new leaves will form.

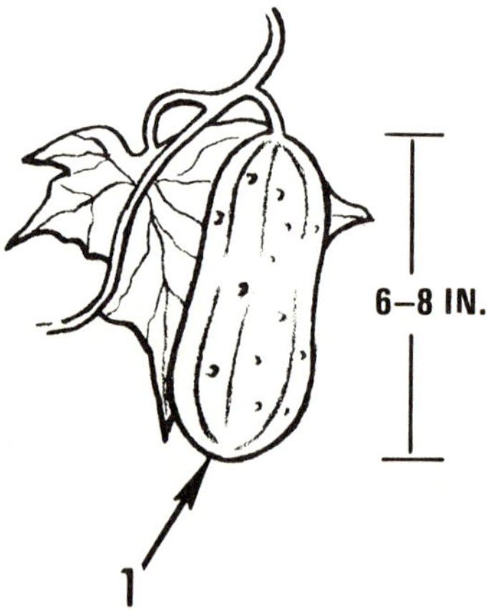

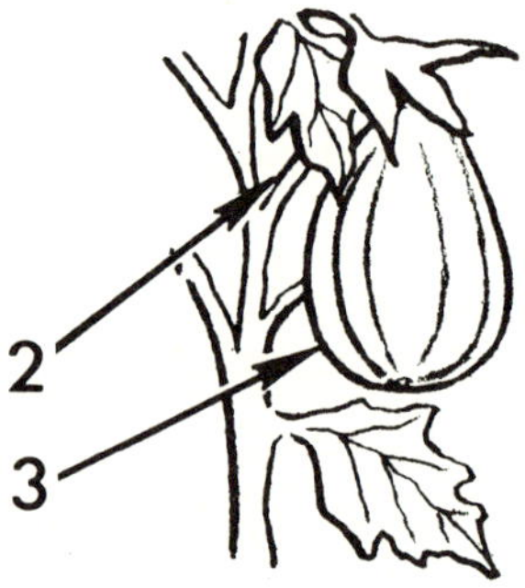

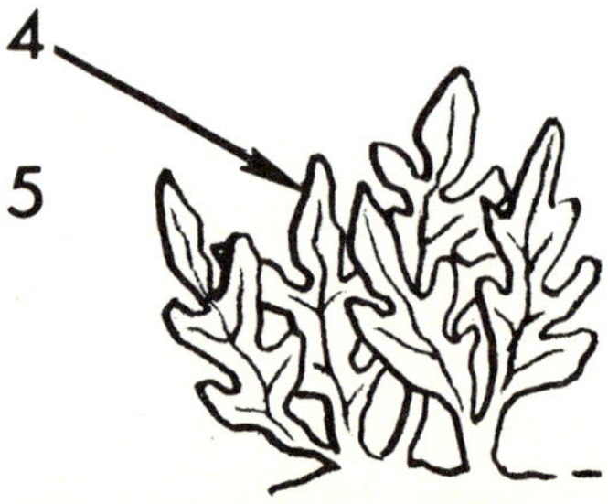

Harvesting

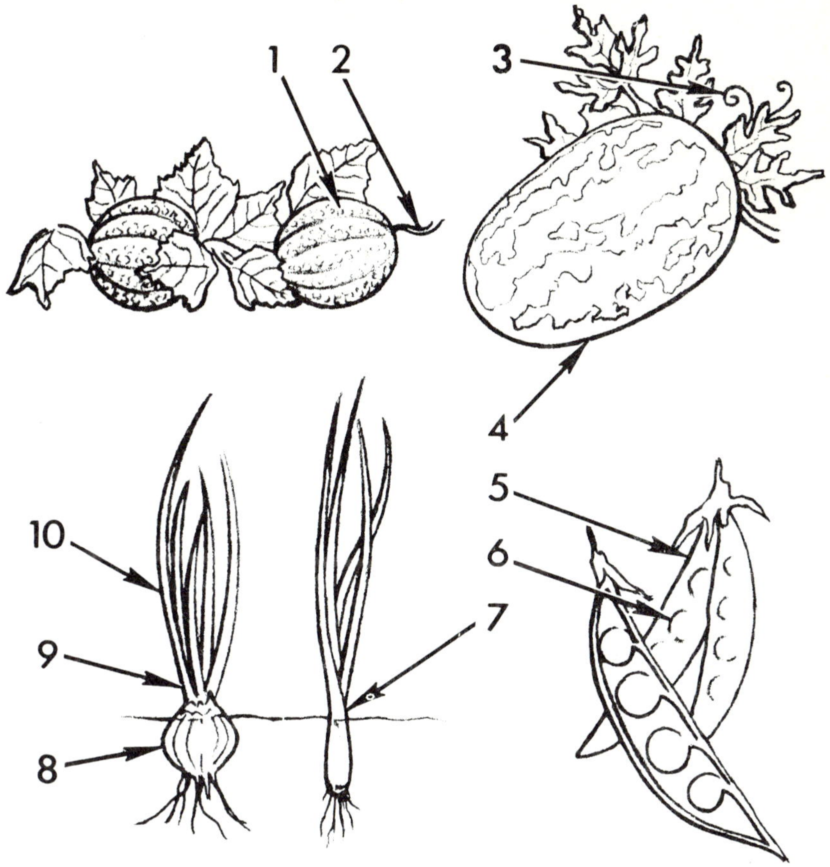

Melons: Cantaloupes are ready to eat when the netting [1] on the skin becomes thick and corky, and the areas between the netting have turned from green to tan. The stem [2] should pull off very easily. Honeydew and casaba melons are ripe when the skin is completely yellow.

Watermelons [4] are more difficult to judge. In the morning, rap with your knuckles. There should be a dull sound, not a "ping". Light areas on the underside of the fruit should turn from white to yellow. The two curly tendrils [3] on the stem nearest the fruit should turn to brown.

Onions: Tops [10] of bulb onions will dry and become yellow. When tops are dry, dig bulbs [8] and dry for a few days. Cut stem [9] leaving 2 to 3 inches attached.

Green onions should be pulled as needed when the stems [7] are about 1/4 inch thick.

Peas: Pods [5] should be picked when rounded pea shapes [6] become visible. Keep picking mature pods, to increase growth of new pods. Like corn, peas start losing flavor after picking. Therefore, plan to pick just before cooking.

Harvesting

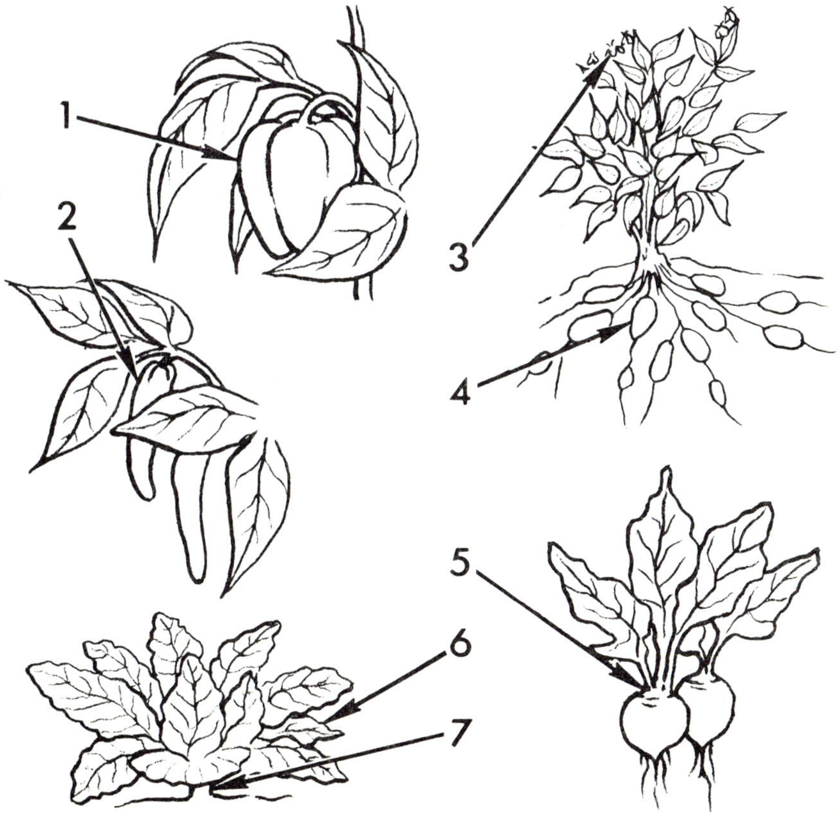

Peppers: Pick sweet peppers [1] while still green, firm, and crisp. Cut the stem, don't pull. If left on plant, peppers turn red and sweeter.

Hot peppers [2] should be a bright red before picking.

Potatoes: Potatoes [4] are ready to be dug up when the tops [3] begin to dry and bend over.

Radishes: A small red crown [5] appears just above the ground when radishes are ready for harvest. Smaller radishes are quite tender. Radishes become pithy as they get larger.

Spinach: Cut the outer leaves [6] near the base [7] when the leaves are 3 inches long or longer. The inner leaves are then cut later. Or, cut off the entire plant near the base [7]. Leaves will not grow again.

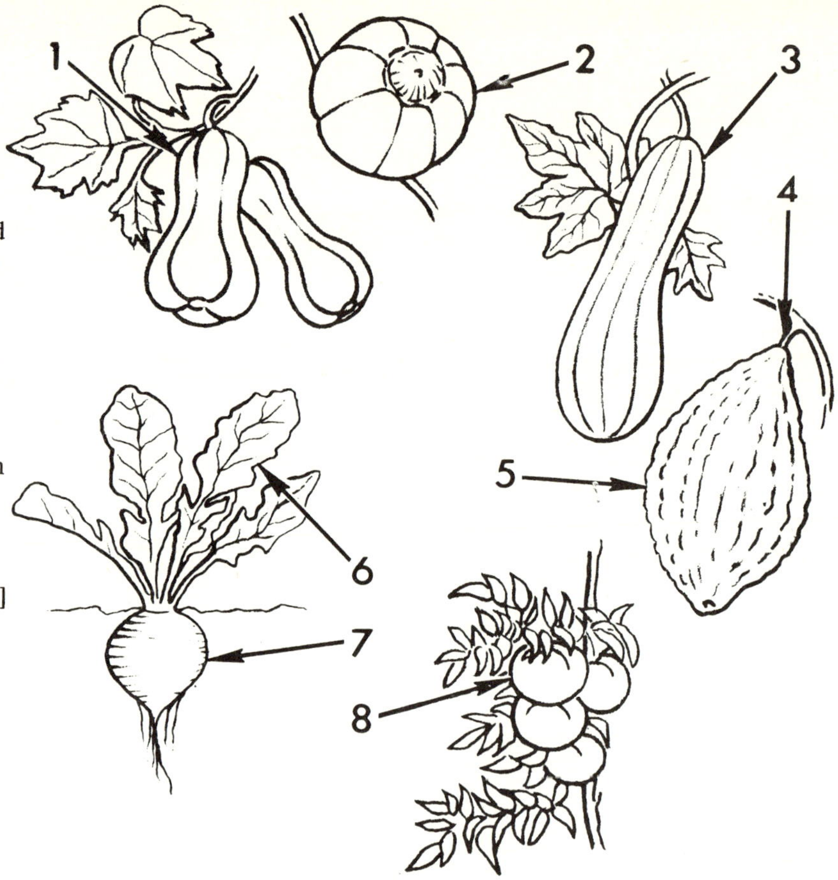

Harvesting

Squash: Summer squash [1,2,3] should be picked while still fairly small. The skin should be tender, and easily pierced with a fingernail. Keep picking to encourage growth of new squash.

Winter squash [5] should be left on the vine until the skin is very hard. The stem [4] turns light yellowish green. Cut the stem, do not pull.

Tomatoes: Tomatoes [8] are best harvested when they reach their full red color. However, they may be picked earlier and stored for ripening in a warm, dark place.

Turnips: Turnips are harvested when the roots [7] are 2 to 4 inches in diameter. When they are larger, they become pithy. The leaves [6] can be harvested for turnip greens when the turnips are young.

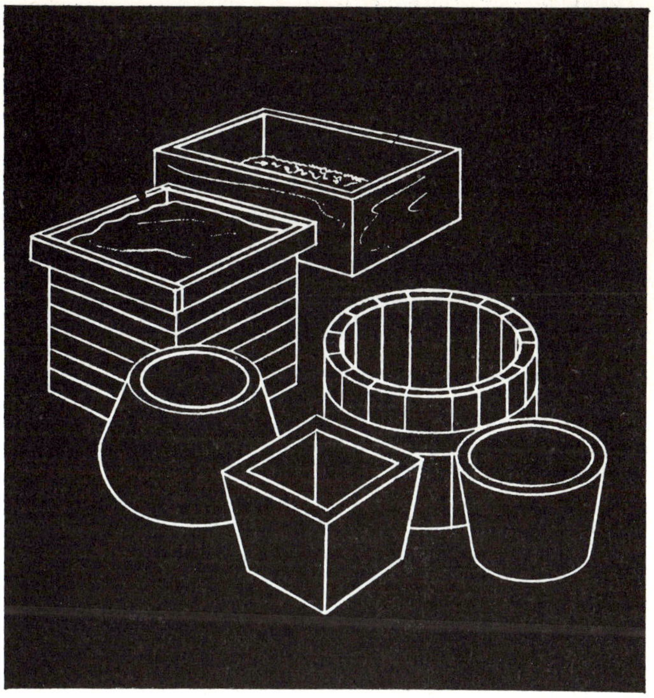

GROWING VEGETABLES IN CONTAINERS

WHY CONTAINER PLANTING IS USEFUL

Growing vegetables in containers can solve some problems of the home vegetable gardener. First, some potential gardeners have no space available for a conventional garden. This is the case with many occupants of apartments, condominiums, and townhouses.

Second, even if there is ground space available, the space may be unsuitable for a vegetable garden. There may not be enough sunlight, drainage may be poor, or the soil may be so bad as to require replacement.

Finally, container planting has some positive benefits of its own. The home gardener may choose to plant some vegetables in containers even though not driven to do so by necessity. Vegetables in containers can make an attractive display and be used for decoration. Because containers are portable, the display can be easily changed periodically as the vegetables reach maturity and are harvested. Portability is also an advantage from another standpoint: the containers can be moved into the garage or other roofed area if there is danger of frost.

Container gardening, as we will see, is not difficult, and can be very productive and satisfying. Many types of containers and many special varieties of vegetables have been developed to help the container gardener.

SELECTING VEGETABLES FOR CONTAINER PLANTING

Much of the discussion on Pages 19 and 20 for selecting vegetables for the standard vegetable garden applies also to growing vegetables in containers. In particular, you will want to select vegetables which will grow well in your climate and which you enjoy eating.

However, not all of the 24 vegetables described earlier are suitable for container planting. While it is true that every one of them could be grown in containers, some vegetables require such large containers or produce such low yields that they are less desirable for limited-space planting.

These less-desirable vegetables for container planting are:

asparagus	–	large root system [1]
broccoli	–	large plant [2]
cauliflower	–	large plant [3]
celery	–	deep roots [4], needs lots of water and fertilizer over long period
melons	–	large plant [5], low yield, slow growth
peas	–	deep roots [6], low yield
potatoes	–	large root system [7]
squash	–	large plant [8]

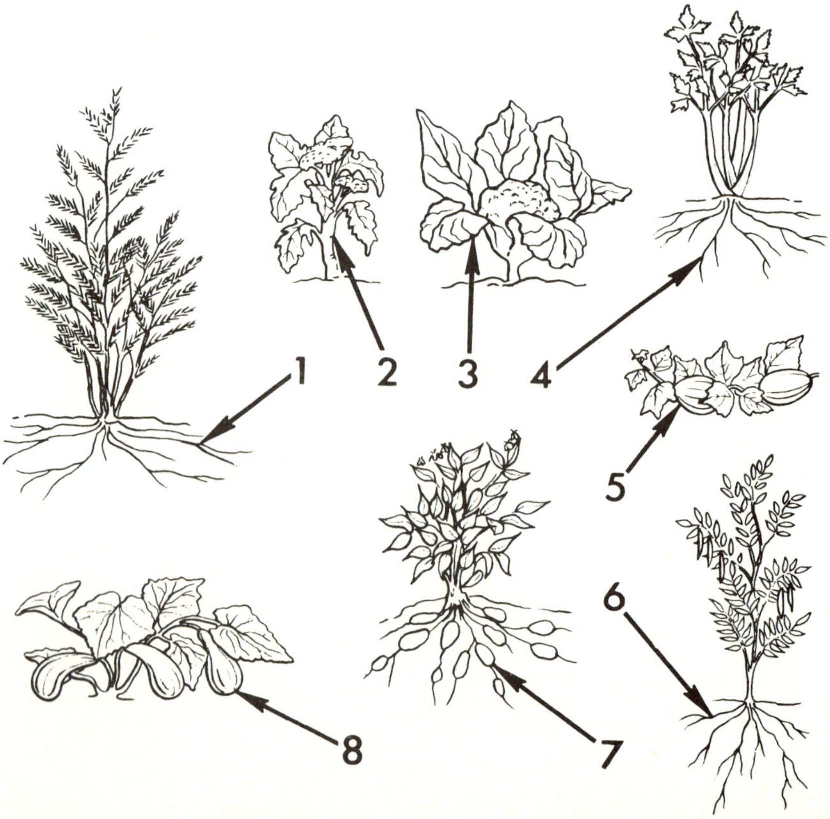

Vegetables which are good to excellent for container planting are:

beans	–	bush beans are more compact than pole beans and therefore easier to manage in containers
beets	–	excellent for container planting
brussels sprouts	–	the plant is large but very productive
cabbage	–	the normal plant is large, with deep roots—not suitable for container planting. However, miniature varieties are available
carrots	–	soil depth is required but short varieties and miniatures are available
chard	–	excellent for container planting, heavy continuous yield from small space
corn	–	not the best for containers because of space requirements and low productivity. However, midget varieties are available which are more suitable
cucumbers	–	normal varieties require lots of space. Like corn, smaller varieties are available for container planting
eggplant	–	excellent for container planting, attractive plant, good yield. Also, small varieties are available
lettuce	–	excellent for container planting
onions	–	excellent for container planting. Green onions are particularly easy to grow
peppers	–	excellent for container planting. Compact, attractive plant, high yield
radishes	–	excellent for container planting. Easy to grow and fast growing.
spinach	–	good for container planting but relatively low yield
tomatoes	–	standard varieties require good size containers because of deep roots. Midget varieties are available which are excellent for container planting
turnips	–	good soil depth required, but productivity is high

▶ Miniature Varieties

There is a wide selection of miniature varieties available which are ideal for container planting. Both the plant and the fruit are smaller in most of these varieties, but the flavor is excellent. Seeds for miniature varieties are available from garden stores, nurseries, and by mail order. A list of major seed companies is provided at the right.

Some of the more popular miniature varieties are:

cabbage	–	Baby Head, Dwarf Modern, Little Leaguer
carrots	–	Baby Finger, Little Finger, Short'n Sweet, Tiny Sweet
corn	–	Golden Midget, Midget, White Midget
cucumbers	–	Little Minnie, Patio Pik, Tiny Dill
eggplant	–	Early Hybrid, Modern Midget, Golden Yellow
tomatoes	–	Dwarf Champion, Patio, Pixie, Presto, Small Fry, Stokes Stakeless, Tiny Tim

MAJOR SEED COMPANIES

1. Burgess Seed & Plant Co.
   P.O. Box 218
   Galesburg, MI 49053
2. W. Atlee Burpee
   Philadelphia, PA 19132
   Clinton, IA 52732
   Riverside, CA 92502
3. Farmer Seed & Nursery
   Faribault, MN 55021
4. Henry Field Seed & Nursery Co.
   Shenandoah, IA 51601
5. Gurney Seed & Nursery Co.
   Yankton, SD 57078
6. Joseph Harris Co., Inc.
   Moreton Farm
   Rochester, NY 14642
7. J.W. Jung Co.
   Randolph, WI 53956
8. Earl May Seed & Nursery Co.
   Shenandoah, IA 51601
9. R.H. Shumway
   P.O. Box 777
   Rockford, IL 61101

▶ Herbs

Herbs are easy to grow and are ideal for container planting. They can be used fresh for cooking, or they can be dried for storage (see Drying Herbs, Page 81). Some of the common herbs are as follows:

basil — the leaves are used to flavor tomato, fish, cheese and many other dishes. Basil is very common in Italian foods. The plant is an attractive annual.

chives — the long, hollow leaves are clipped when needed for flavoring salads, soups, and cheese and egg dishes. The plants are perennial, and produce decorative flowers.

dill — the seeds and leaves are used for flavoring fish, lamb, for pickling, and in salads. The plant grows quite large, to a height of 3 to 4 feet. It is an annual which usually reseeds itself.

marjoram — the small leaves are used as a seasoning for soups, stews, beef, lamb, and in Italian dishes. The plant is a perennial if protected in the winter or if the winters are mild. Otherwise, it is grown as an annual.

BASIL

CHIVES

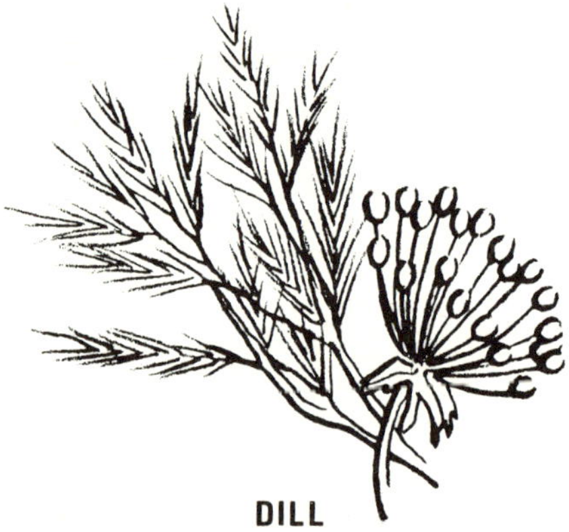
DILL

MARJORAM

Herbs

mint — there are many mint flavors, including spearmint, peppermint, orange mint, apple mint, and dozens of others. Mint is used for cool drinks such as iced tea, as well as for lamb, peas and carrots, and other dishes. Mint is a perennial.

oregano — the leaves are picked as needed for seasoning many dishes, particularly Italian and Mexican food. The plant is a perennial.

parsley — the stems and leaves are used primarily as a garnish, or are chopped and used in soups, stews, and salads. The plant is a biennial and will live to the next year in mild winters.

rosemary — the leaves are used for seasoning fish, lamb, pork, veal, and in many sauces. The plant can reach a height of 5 to 6 feet, unless clipped back. The plant is a perennial, if protected over the winter.

MINT

OREGANO

PARSLEY

ROSEMARY

Herbs

sage — the leaves are used in dressings and stuffing for poultry and pork, and in making sausage. The plant is a perennial, with many varieties.

savory — the leaves are picked during the growing season for flavoring bean and pea dishes, eggs, salads and soups. There are two kinds of savory, summer and winter. Summer savory is the more common, and is an annual. Winter savory is a perennial.

tarragon — the leaves are used in salad dressings, sauces, stews, and for seasoning fish, chicken, and many other dishes. There are two types of tarragon, French and Russian. The French is considered far more flavorful. It is a perennial.

thyme — there are many kinds of thyme—lemon thyme, golden thyme, caraway thyme, etc. All are used in meat sauces, ground beef and pork, and in soups and vegetable dishes. All thymes are perennials.

SAGE

SAVORY

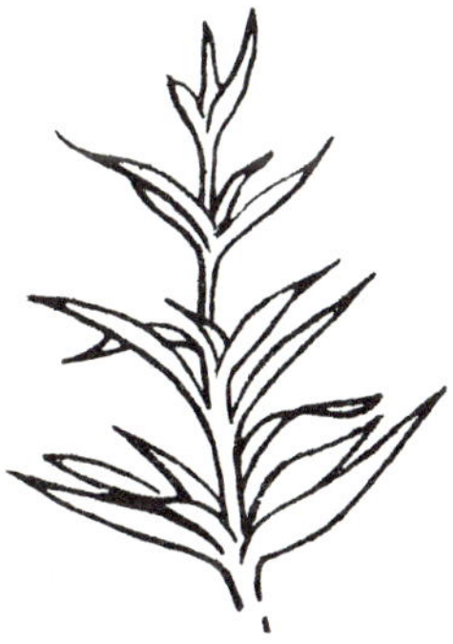

TARRAGON

THYME

PLANTING AND CARE

▶ Types of Containers

There are few restrictions in selecting containers for planting vegetables. The containers must be strong enough to hold planter mix (which is quite light) and be able to withstand frequent watering. Appearance of the containers is a matter of personal preference.

The containers must also be large enough. The size depends on what vegetables are to be planted. Radishes and lettuce require a soil depth of only about 6 inches. This, combined with room at the bottom for gravel (see Page 50) and some room at the top, means a container depth of about 8 inches will be satisfactory. Most vegetables need about 12 inches of soil, so that containers should be at least 14 inches in depth. It is better to have deeper containers to improve drainage. In addition to crowding the roots, small containers tend to dry out fast, making frequent watering necessary. Also, small containers tip over easily.

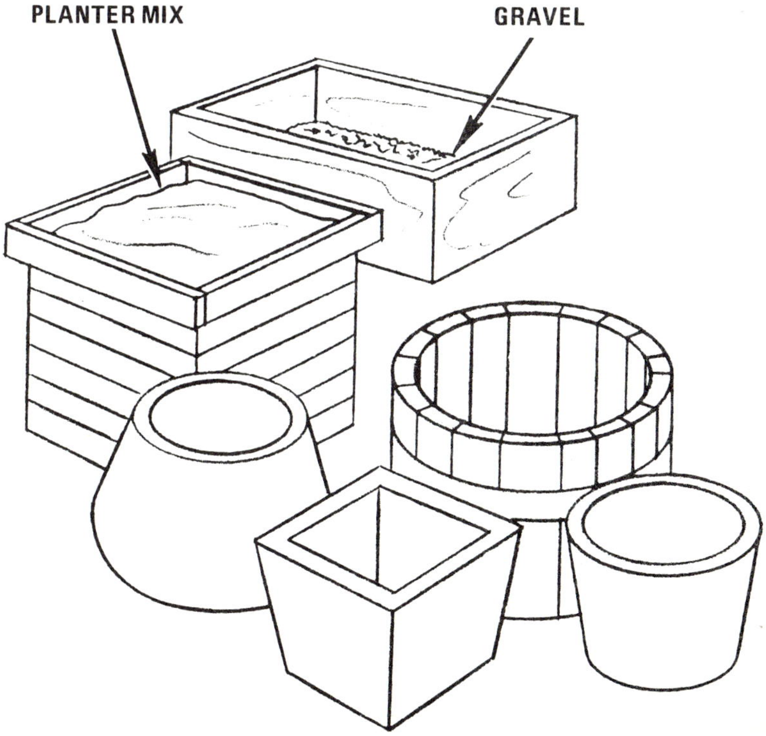

Thus, many types of containers will serve as vegetable planters. There are many sizes and shapes of clay and ceramic pots available, which are sometimes expensive. Wooden boxes, redwood planters, pressed-paper tubs, plastic trash buckets, bushel baskets, and many others can serve as containers. Containers can be either bought or you can find or make your own.

PLANTING AND CARE

Types of Containers

Containers must have good drainage. For small containers (1 foot or less on a side), drill 4 holes in the bottom or near the bottom. Make the holes at least 1/4 inch in diameter, and space them evenly around the container.

For larger containers, place one hole [1] every foot around the container.

If the holes are drilled near the bottom, make the holes about 1/2 inch above the inside bottom of the container. This is to insure that the holes will be covered by gravel [3] rather than planter mix [2].

If the container is fragile and does not appear able to support the soil and water, it can be lined with a plastic garbage bag. However, if a bag is used, drainage holes must be cut next to the holes in the container.

One advantage to wooden or pressed-paper containers over clay or ceramic is that wooden containers or those made of pressed paper can survive the winter without cracking or breaking if the soil is left in them. However, except in areas with mild winters, clay or ceramic containers should be emptied of soil in preparation for winter.

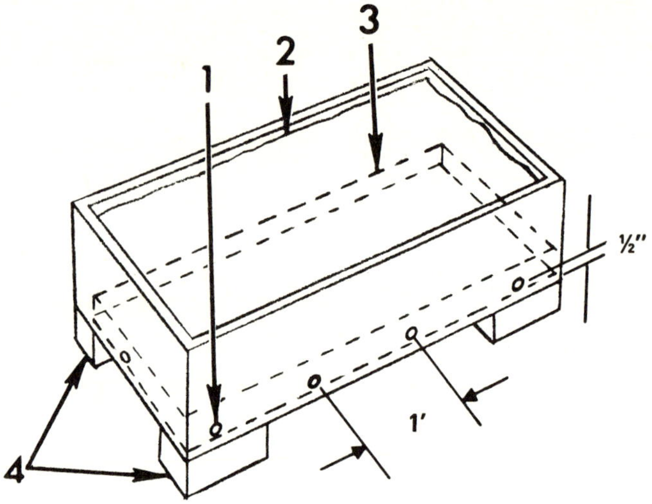

Often containers are placed on a paved area, such as a patio. Such areas can become quite warm from the sun's heat. In turn, the containers can become overly warm. If you live in a warm climate, try placing bricks [4] or wooden blocks under your containers to allow air to circulate.

▶ The Planting Mix

Ordinary soil is not good for container planting. Ordinary soil tends to be too heavy, and does not absorb or drain water fast enough. Also, ordinary soil is inclined to form a heavy mass which shrinks away from the sides of the container. Then, when you water, the water runs down the sides and out, rather than reaching the roots.

It is better to use a planter mix such as those available widely in nurseries and garden stores. Or, you can make your own mix from the following materials suggested by the U.S. Department of Agriculture:

- 1 bushel vermiculite (about 1-1/4 cubic feet)
- 1 bushel peat moss (about 1-1/4 cubic feet)
- 1-1/4 cups ground limestone
- 1/2 cup 20% superphosphate
- 1 cup 5-10-5 fertilizer

These materials are then mixed thoroughly, and will produce about 2-1/2 cubic feet of planting mix. The mix is light and porous, and will have good absorption and drainage properties.

Before pouring the planting mix [1] in the containers, place about 1 inch of coarse gravel [2] or rock in the bottom of each container. This will assist drainage and prevent the planter mix from gradually seeping out the drainage holes [3].

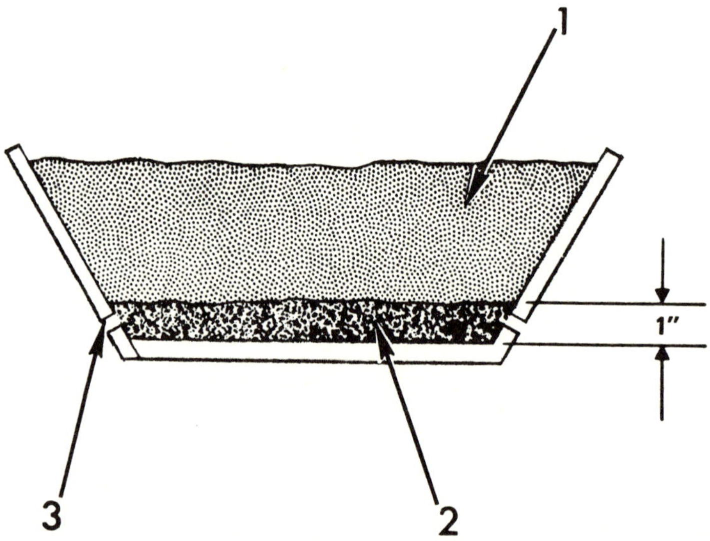

▶ When to Plant

Planting in containers does not differ from planting in open ground for such basic factors as determining when to plant. These factors are discussed in detail on Pages 26 to 28, and the major guidelines discussed there apply to container planting as well.

The most important factor is whether the individual vegetable is a cool-season or warm-season vegetable. This will guide your decision as to whether to grow the vegetable in your climate and if so, when to plant it.

Succession planting of different types of crops—cool season, followed by warm season, followed by cool season—applies also to container planting. Similarly, succession planting of the same crop, to stretch out the harvest season, also applies. However, instead of dividing an individual container into separate parts for succession planting, you may wish to provide additional containers for this purpose.

Starting seeds indoors to produce your own seedlings for transplant is also just as appropriate for planting in containers. The time of starting such seedlings depends on the climate and the vegetable, not on whether you will be planting in containers.

Herbs

The time to plant herbs, however, has not yet been discussed.

Herbs are planted in different ways, as described later (see How to Plant, below). In general, annuals are planted from seeds planted outdoors in the containers in spring, after the last frost. These annuals are: basil, dill, parsley, and summer savory. (Parsley is actually a biennial, but is treated as an annual.)

Perennials are grown from seedlings or cuttings. You can grow your own seedlings by starting seeds indoors in winter. The seedlings or cuttings are then transplanted outdoors in the containers after the last frost in spring. These perennials are: chives, marjoram, mint, oregano, rosemary, sage, winter savory, tarragon, and thyme.

▶ How to Plant

Planting in containers is similar to planting in open ground (see Pages 28 to 31).

Seed is broadcast and covered lightly with planting mix, or is planted in small rows. The edge of a board [2] is a convenient tool for making straight rows [1] of uniform depth. After planting seed [3], cover lightly with planting mix [4]. Keep planting mix moist until germination.

Plants or seedlings [5] are planted carefully in the planting mix [4], being careful not to disturb the roots. The seedlings should be held by the leaves, not the stem.

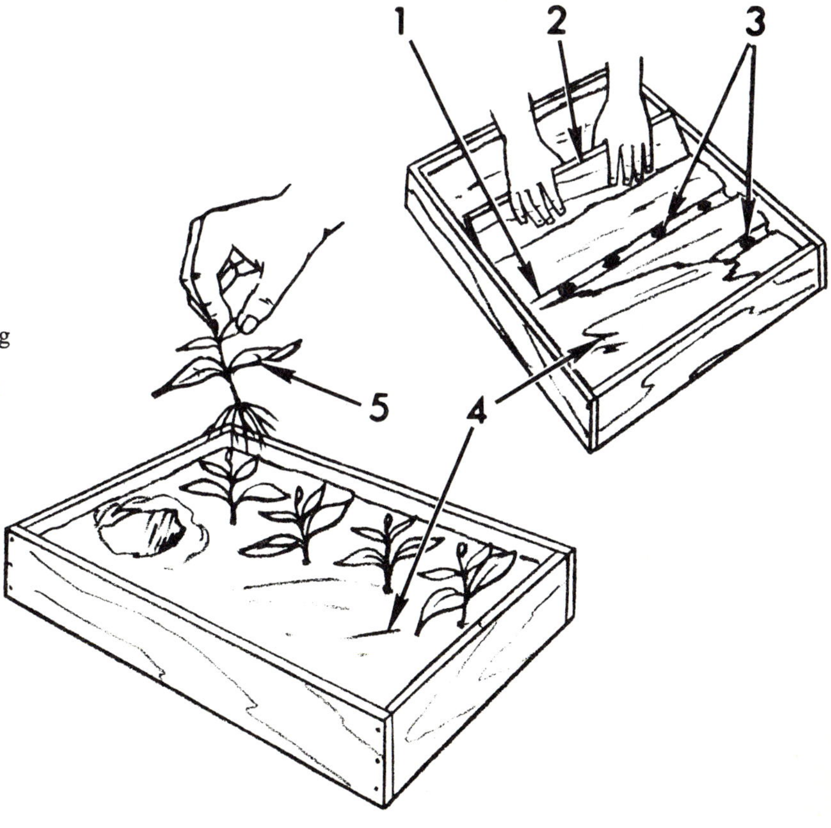

How to Plant

Herbs

Some herbs are planted as seed directly in the containers, some are transplanted, some are planted as cuttings, and some in all three ways. Those herbs which are transplanted can either be grown from plants which you buy or from seedlings which you grow yourself from seed.

The accompanying table shows how the different herbs can be planted. If an "X" appears in more than one column, that means the herbs can be planted in the different ways indicated. For example, tarragon can be grown from plants bought at the nursery or from cuttings taken from a growing plant.

	Plant Seeds Directly in Container	Grow Seedlings Indoors and Transplant to Container	Buy Plants and Transplant to Container	Plant Cuttings Taken from Growing Plant
Basil	X			
Chives			X	
Dill	X			
Marjoram		X	X	
Mint			X	X
Oregano		X	X	
Parsley	X*			
Rosemary		X	X	X
Sage		X	X	X
Savory				
Summer	X			
Winter		X	X	X
Tarragon			X	X
Thyme		X	X	X

*Soak seeds for 24 hours in warm water before planting

How to Plant

How to plant cuttings needs to be explained. It is very easy to do. During spring or early summer, which is the period of active growth, cut one or more of the side shoots [1]. Place them firmly in the planting mix and water carefully. Roots will form, usually in a few weeks, and you will have a new plant.

The cuttings can be planted in individual peat pots [2] if desired. This reduces the chance of root damage.

The rooting process can be speeded up by adding root hormone at the base of the cutting. Follow directions on the label.

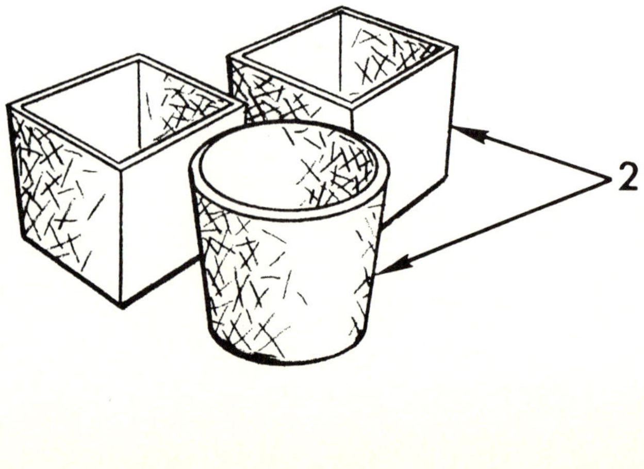

▶ Watering and Feeding

Watering is important for vegetables planted in containers. Plants in containers cannot develop deep root systems, so more frequent watering is necessary.

Check the planting mix daily for moisture. If it is dry just beneath the surface, water until the water comes out of the drain holes [2]. A bubbler [3] is useful to prevent splashing. Or wrap a piece of gunny sack around the end of the hose. In hot weather, it may be necessary to water as often as once a day.

Because of the frequent watering required, frequent feeding with a fertilizer is needed. The watering washes out the nutrients, which must be replaced. Plan to fertilize about every 2 weeks, using a liquid fertilizer such as fish emulsion [1] or a solid fertilizer such as 5-10-5 [4]. The fish emulsion should be diluted and applied with a watering can [5]. The solid fertilizer should be mixed into the top 1/2 inch of planter mix. Follow directions on the label regarding amounts.

However, only one fertilizing is needed if you use "Precise" Timed Release Vegetable Food 9-12-6 [6] or Tomato Food 6-18-6. Apply 2 tablespoons per square foot of soil surface. Nutrients are released each time the container is watered.

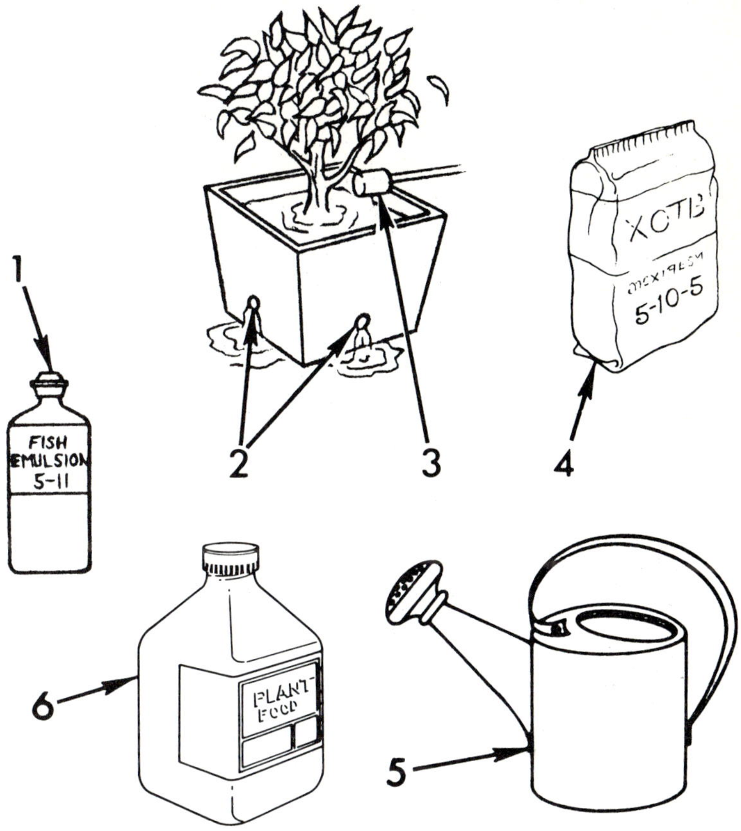

PRESERVING VEGETABLES

There are four basic methods of preserving vegetables once they are harvested: freezing, canning, drying, and cold storage.

Freezing

Freezing preserves the color, texture, and taste of vegetables quite well. However, storage space for frozen vegetables is expensive and therefore is usually quite limited. A separate freezer is desirable, as the freezer compartments of most refrigerators are not designed for lengthy storage of frozen vegetables. They are small and are opened and closed frequently. Both upright [1] and chest [2] freezers are available.

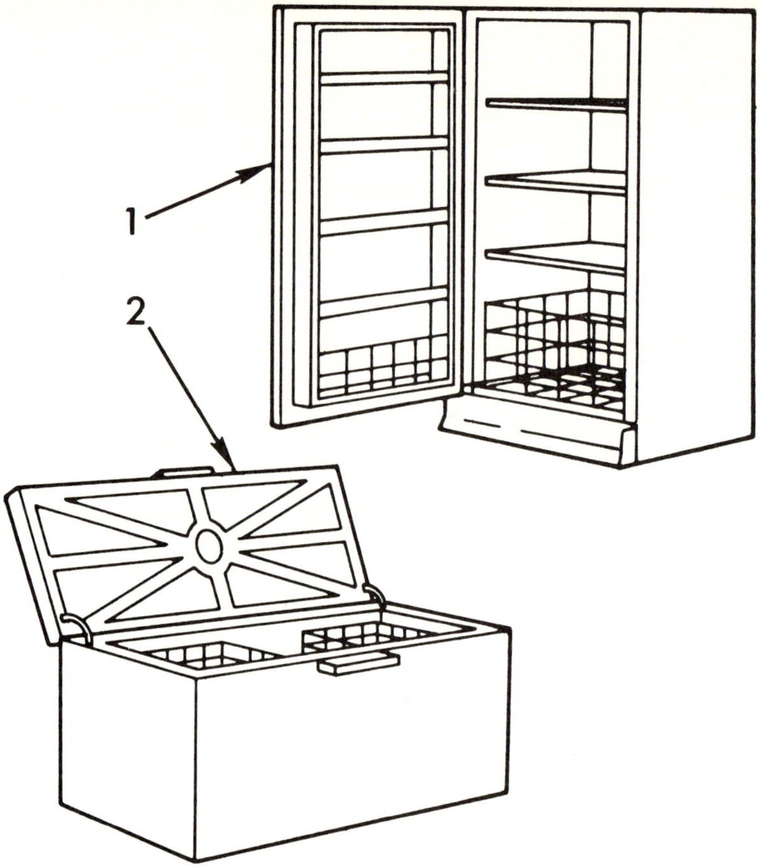

Freezing works well with vegetables which are normally cooked before eating: asparagus, beans, beets, broccoli, brussels sprouts, carrots, cauliflower, corn, eggplant, peas, peppers, potatoes, spinach, squash, turnips.

Freezing is not as effective for vegetables which are normally eaten raw: celery, cucumbers, lettuce, onions, radishes, tomatoes. Cabbage, which is eaten both raw and cooked, does not freeze too well. Tomatoes can be frozen satisfactorily if cooked beforehand (as described later, Page 65).

Canning

Canning requires the purchase of special equipment, most of which can be reused. This includes a pressure canner [1] and jars [2]. Storage space is unlimited, provided containers are available.

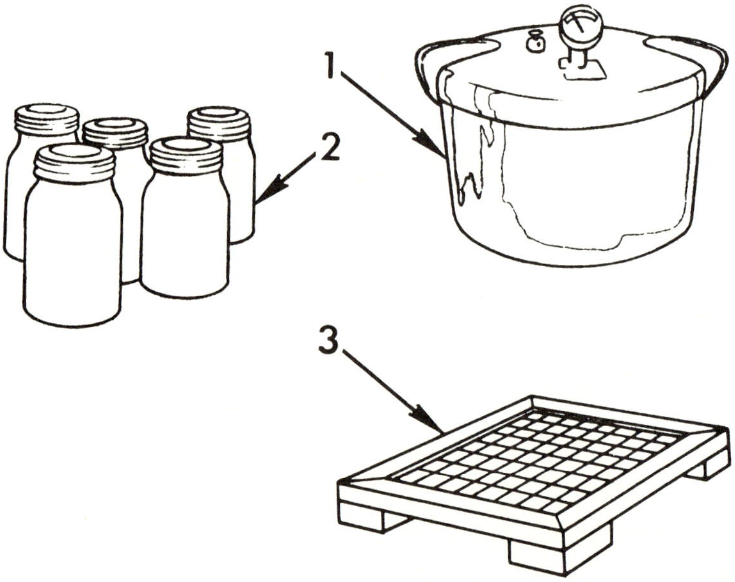

Vegetables which are good for canning are: asparagus, beans, beets, carrots, corn, peas, spinach, squash, tomatoes.

Drying

Drying is one of the oldest methods of preserving foods, and one of the simplest. Water is removed, and nothing is added. A simple drying rack [3] can be home made, as described later (Page 78).

Although most vegetables can be preserved by drying, many show an appreciable loss in color and taste. Herbs, however, are ideal for drying.

Cold Storage

Many vegetables store well in a cold or cool and moist atmosphere. Such an environment often exists in cellars or basements, or can be created by such means as underground storage.

All vegetables can be stored for at least a few days or weeks in the proper conditions. The root crops are particularly good for storage: beets, carrots, onions, potatoes, radishes, turnips.

Vegetables should be prepared and frozen, canned, dried, or stored as quickly as possible after harvest. This will increase the storage life and eating quality of the vegetable.

If for some reason you cannot prepare and store the vegetable right after picking, cool the vegetable by immersing it in ice water. Do not keep it at room temperature or expose it to the sun.

After cooling the vegetable in ice water, drain it and keep it at a low temperature (32°–40°F). Do not wait longer than 24 hours before preparing and storing the vegetable.

▶ Varieties of Vegetables

Certain varieties of vegetables have been developed which are especially good for freezing, canning, or storing. These varieties are listed later in the individual sections on freezing, canning, and storing. The seed companies from which each variety can be bought are also indicated there.

As a convenient summary, the accompanying table shows the varieties which are good for freezing, canning, and storing. You will notice that many varieties are good for both freezing and canning.

Varieties of Vegetables

GOOD VEGETABLE VARIETIES FOR FREEZING, CANNING, OR STORING

Vegetable	Varieties	Variety is Good For: Freezing	Canning	Storing
Asparagus	California 500	X		
	Mary Washington	X		
	Roberts	X		
Beans				
Snap				
Bush	Blue Lake	X	X	
	Brittle Wax	X	X	
	Burpee's Stringless	X		
	Butter Wax		X	
	Cherokee Wax	X		
	Contender	X	X	
	Garden Green	X	X	
	Golden Wax	X		
	Greencrop	X		
	Green Isle	X	X	
	Improved Tendergreen	X	X	
	Kinghorn Wax	X	X	
	Leka Lake	X	X	
	Pencil Pod Wax	X		
	Royalty Purple	X		
	Spartan Arrow	X		
	Tenderette	X	X	
	Top Crop	X	X	
	Wade	X		
Pole	Blue Lake	X	X	
	Kentucky Wonder	X		
	New Pole	X	X	
	Purple Pod		X	

Varieties of Vegetables

GOOD VEGETABLE VARIETIES FOR FREEZING, CANNING, OR STORING (contd)

Vegetable	Varieties	Variety is Good For: Freezing	Canning	Storing
Beans (contd)				
Lima	Baby Bush	X	X	
	Burpee's Best	X		
	Clark's Green Seeded	X	X	
	Fordhook 242	X	X	
	Henderson's	X	X	
	Improved Giant Bush	X		
Beets	Detroit Dark Red			X
	Early Wonder		X	
	Pacemaker		X	
	Red Ball	X	X	
	Red Ruby Queen		X	
	Sweetheart	X	X	
Broccoli	Greenbud	X		
	Green Sprouting	X		
	New Spartan	X		
Brussels sprouts	Jade Cross Hybrid	X		
Cabbage	Autumn King			X
	Copenhagen Market			X
	Danish Bullhead			X
	Early Flat Dutch			X
	Jumbo			X
	Mammoth Red Rock			X
	Penn State Bullhead			X
	Premium Late Flat Dutch			X
	Surehead			X
	Wisconsin All Season			X
	Wisconsin Red Hollander			X
Carrots	Coreless	X		
	Danvers Half Long			X
	Gold Pak	X	X	X
	Nantes Half Long	X	X	
	Red Cored Chantenay	X		
	Royal Chantenay	X		
	Tendersweet	X		
Cauliflower	Early Snowball	X		
	Purple Head	X		
	Super Snowball	X		
Celery	Fordhook			X
	Giant Pascal			X
	Golden Self-Blanching			X
Chard	Lucullus	X		
Corn	Butter Nugget	X		
	Carmel Cross Hybrid	X		
	Country Gentleman Hybrid		X	
	Early Golden Main Crop	X	X	
	Early Sunglow Hybrid	X		
	Golden Bantam	X		
	Golden Beauty	X	X	
	Golden Cross Bantam	X	X	
	Golden Delicious	X		
	Hybrid Truckers	X		
	Iochief		X	
	Jubilee	X	X	
	Kanner King	X	X	
	Marcross Hybrid	X	X	

Varieties of Vegetables

GOOD VEGETABLE VARIETIES FOR FREEZING, CANNING, OR STORING (contd)

Vegetable	Varieties	Variety is Good For:		
		Freezing	Canning	Storing
Corn (contd)	Most Tender	X	X	
	New Cheddar Cross Hybrid	X	X	
	Six Shooter Sugar Corn	X		
	Stowell's Evergreen	X	X	
	White Evergreen		X	
	Wonderful	X		
	Xtra Sweet	X		
Onions, bulb	Downing Yellow Globe			X
	Ebenezer			X
	Hybrid Yellow Sweet Spanish			X
	Red Wethersfield Hamburger			X
	Ringmaster			X
	Southport Red, White and Yellow Globes			X
	White Sweet Keeper			X
	White Sweet Slicer			X
	White Sweet Spanish			X
Peas	Alaska		X	
	Blue Bantam	X		
	Burpeanna Early	X		
	Early All-Sweet		X	
	Freezonian	X		
	Frosty	X		
	Giant Stride	X	X	
	Green Arrow	X		
	Laxton's Progress	X	X	
	Lincoln	X		
	Little Marvel	X	X	
	Miragreen	X		
	Perfected Freezer	X		
	Progress No. 9	X		
	Sweet Green	X		
	Tall Telephone	X		
	Thomas Laxton	X		
	Victory Freezer	X	X	
	Wando	X	X	
Peppers				
Sweet	Early Thick Meat	X		
	Sunnybrook		X	
	Sweet Chocolate	X		
	Tasty		X	
	Worldbeater		X	
Hot	Hungarian Wax		X	
	Pimiento		X	
Potatoes	Anoka	X		
	Kennebec			X
	Norgold Russet			X
	Red Pontiac			X
	White Cobbler			X
Radishes	Cherry Belle			X
	China Rose			X
	Comet			X
	Long Black Spanish			X
	Round Black Spanish			X
	Round White			X
	Stop Lite			X
	White Chinese			X
	White Giant			X

GOOD VEGETABLE VARIETIES FOR FREEZING, CANNING, OR STORING (contd)

Vegetable	Varieties	Variety is Good For:		
		Freezing	Canning	Storing
Spinach	America	X		
	Bloomsdale	X		
	Giant Thick Leaf	X	X	
	Hybrid No. 7	X	X	
	King of Denmark	X		
	New Hybrid	X	X	
	Northland	X	X	
	Virginia Blight Resistant	X	X	
Squash				
Summer	Prolific Straightneck	X		
Winter	Buttercup	X		X
	Butternut	X		
	Gold Nugget	X		
	Golden Delicious	X	X	
	Hubbard (Golden or Red)	X	X	X
	Hubbard (True or Warted)			X
	Royal Acorn			X
	Table Queen			X
	Waltham Butternut			X
Tomatoes	Beefsteak		X	
	Colossal		X	
	Droplet	X*		
	Crimson Giant		X	
	Garden State		X	
	Halcan Canner		X	
	Little Peas	X*		
	Marglobe		X	
	Pink Gourmet		X	
	Pinkshipper		X	
	Ponderosa		X	
	Red Pear	X*	X	
	Red Sugar	X*	X	
	Roma		X	
	Rutgers		X	
	Tiny Tim Midget	X*		
	Yellow Husk	X*		
	Yellow Pear	X*	X	
Turnips	Purple Top White Globe			X

*Frozen stewed tomatoes

Freezing of vegetables should be performed as soon as possible after harvesting. For this reason, it is well to have all equipment and containers clean and ready before you start.

Freezing directions will be given for the following vegetables: asparagus, beans, beets, broccoli, brussels sprouts, cabbage, carrots, cauliflower, celery, chard, corn, peas, peppers (sweet), spinach, squash, tomatoes (stewed), and turnips. Generally, however, beets are preferred canned. Carrots are more commonly preserved in cold storage. Celery that is frozen should be used for cooking, such as in casseroles, stews, and soups.

▶ Freezing Varieties

Different varieties of vegetables have been developed which are especially good for freezing. Some of these varieties are shown in the accompanying table on the next few pages.

The right hand columns of the table show where the different varieties can be purchased: Burgess, Burpee, etc. The full name and address of these seed companies is given to the right.

If an "X" appears under a company name for a particular variety, you can order that variety by writing to the address given.

For example, to order Blue Lake variety snap beans, bush type, you can write to Field, Gurney, Jung, or May.

1. Burgess Seed & Plant Co.
   P.O. Box 218
   Galesburg, MI 49053
2. W. Atlee Burpee
   Philadelphia, PA 19132
   Clinton, IA 52732
   Riverside, CA 92502
3. Farmer Seed & Nursery
   Faribault, MN 55021
4. Henry Field Seed & Nursery Co.
   Shenandoah, IA 51601
5. Gurney Seed & Nursery Co.
   Yankton, SD 57078
6. Joseph Harris Co., Inc.
   Moreton Farm
   Rochester, NY 14642
7. J.W. Jung Co.
   Randolph, WI 53956
8. Earl May Seed & Nursery Co.
   Shenandoah, IA 51601
9. R.H. Shumway
   P.O. Box 777
   Rockford, IL 61101

Freezing Varieties

GOOD VEGETABLE VARIETIES FOR FREEZING

Vegetable	Varieties	Burgess	Burpee	Farmer	Field	Gurney	Harris	Jung	May	Shumway
Asparagus	California 500				X					
	Mary Washington	X	X	X		X	X			X
	Roberts					X			X	
Beans										
Snap										
Bush	Blue Lake				X	X		X	X	
	Brittle Wax			X	X			X		
	Burpee's Stringless		X		X	X				X
	Cherokee Wax					X	X		X	
	Contender					X			X	
	Garden Green					X				
	Golden Wax				X	X				
	Greencrop			X						
	Green Isle			X						
	Improved Tendergreen			X	X	X			X	X
	Kinghorn Wax		X			X	X			
	Leka Lake		X							
	Pencil Pod Wax				X	X		X	X	
	Royalty Purple	X		X						
	Spartan Arrow			X						
	Tenderette				X	X	X			X
	Top Crop	X	X	X		X			X	
	Wade			X						
Pole	Blue Lake				X			X	X	
	Kentucky Wonder		X	X	X	X	X	X	X	X
	New Pole			X						
Lima	Baby Bush		X		X			X		
	Burpee's Best		X							
	Clark's Green Seeded					X				
	Fordhook 242	X	X		X	X	X	X	X	X
	Henderson's		X		X				X	X
	Improved Giant Bush		X						X	X

FREEZING VEGETABLES

Freezing Varieties

GOOD VEGETABLE VARIETIES FOR FREEZING (contd)

Vegetable	Varieties	Burgess	Burpee	Farmer	Field	Gurney	Harris	Jung	May	Shumway
Beets	Red Ball		X							
	Sweetheart			X						
Broccoli	Greenbud		X							
	Green Sprouting	X		X	X	X				
	New Spartan								X	
Brussels sprouts	Jade Cross Hybrid		X			X	X			
Carrots	Coreless					X		X	X	
	Gold Pak			X	X	X		X		X
	Nantes Half Long		X				X			
	Red Cored Chantenay		X	X	X	X		X		
	Royal Chantenay				X		X			
	Tendersweet				X	X			X	
Cauliflower	Early Snowball		X	X				X	X	X
	Purple Head			X			X			
	Super Snowball		X	X				X		
Chard	Lucullus	X	X	X		X				X
Corn	Butter Nugget			X						
	Carmel Cross Hybrid					X				
	Early Golden Maincrop				X					
	Early Sunglow Hybrid					X				
	Golden Bantam		X		X	X			X	X
	Golden Beauty			X						
	Golden Cross Bantam		X	X	X	X				X
	Golden Delicious				X					
	Hybrid Truckers					X				
	Jubilee			X						
	Kanner King			X						
	Marcross Hybrid					X				
	Most Tender				X					
	New Cheddar Cross Hybrid							X		
	Six Shooter Sugar Corn									X
	Stowell's Evergreen		X	X		X			X	X
	Wonderful						X			
	Xtra Sweet	X	X	X	X					
Peas	Blue Bantam			X				X		
	Burpeanna Early		X							
	Freezonian	X	X	X	X		X			X
	Frosty					X	X		X	
	Giant Stride	X								
	Green Arrow		X							
	Laxton's Progress	X		X	X	X				X
	Lincoln			X			X			
	Little Marvel	X	X	X		X			X	
	Miragreen							X		
	Perfected Freezer						X			
	Progress No. 9								X	
	Sweet Green			X					X	
	Tall Telephone			X						
	Thomas Laxton		X			X				X
	Victory Freezer					X				
	Wands	X	X	X	X		X	X		
Peppers										
Sweet	Early Thick Meat								X	
	Sweet Chocolate			X						
	Tasty			X						
Spinach	America								X	
	Bloomsdale		X	X	X			X	X	X
	Giant Thick Leaf				X					
	Hybrid No. 7		X	X						
	King of Denmark								X	
	New Hybrid					X				
	Northland					X				
	Virginia Blight Resistant	X								
Squash										
Summer	Prolific Straightneck	X	X	X		X	X	X	X	X
Winter	Buttercup						X			
	Golden Delicious					X	X			
	Hubbard (Golden or Red)	X	X	X						
Tomatoes										
(stewed)	Droplet			X						
	Little Pear					X				
	Red Pear				X					X
	Red Sugar	X								
	Tiny Tim Midget			X					X	
	Yellow Husk					X				
	Yellow Pear								X	

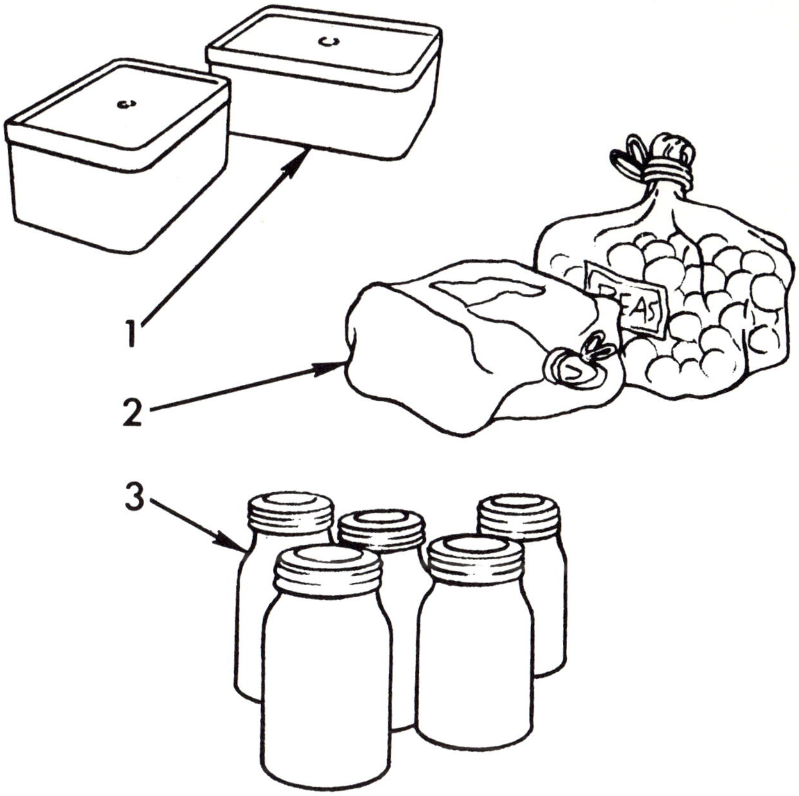

▶ Preparing Equipment

Containers

Any container may be used that will keep out the air and prevent loss of moisture. Common containers for freezing include plastic containers [1], heavy duty plastic bags [2], and glass canning jars [3].

Consider the size and ease of stacking in selecting containers. An ideal size is that which will hold enough vegetable for one meal for your family. Rigid containers that are flat on both top and bottom can be stacked easily and take up less space.

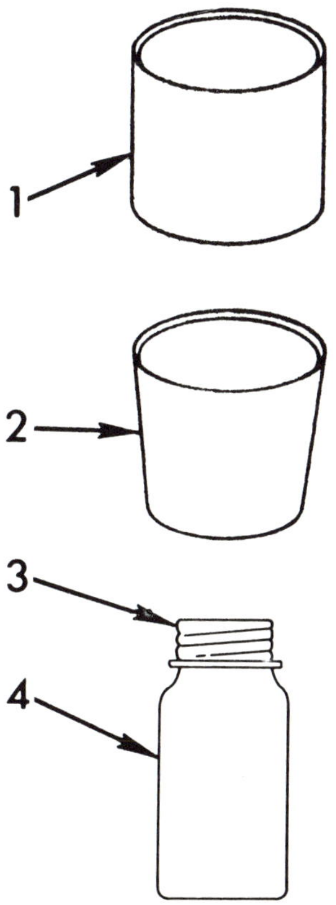

Preparing Equipment

Containers with vertical sides [1], or with sides that flare out [2], are easy to empty of frozen contents. If the opening [3] is narrower than the body [4] of the container the vegetable must be partially thawed before it can be removed.

To assist you in determining your container requirements, the accompanying table shows the approximate amount of fresh vegetable that will produce 1 pint of frozen vegetable.

Amount of Fresh Vegetable to Make 1 Pint of Frozen Vegetable

Vegetable	Pounds
Asparagus	1 – 1½
Beans	
Snap	2/3 – 1
Lima, in pods	2 – 2½
Beets, without tops	1¼ – 1½
Broccoli	1
Brussels sprouts	1
Carrots, without tops	1¼ – 1½
Cauliflower	1⅓
Chard	1 – 1½
Corn, in husks	2 – 2½
Eggplant	1
Peas	2 – 2½
Peppers, sweet or bell	2/3
Spinach	1 – 1½
Squash	
Summer	1 – 1¼
Winter	3

FREEZING VEGETABLES

Preparing Equipment

Blanching Kettle

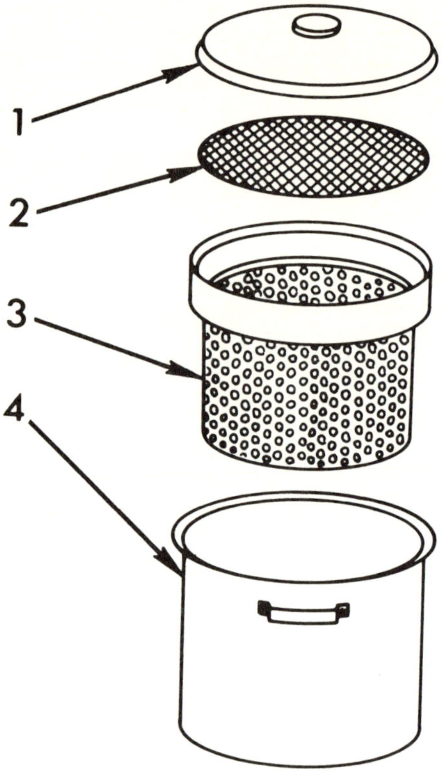

Before freezing, all vegetables should be heated, or "blanched", before packing. This stops the vegetable growth process at the peak of quality and flavor. Without blanching, some growth processes continue in the frozen vegetable, leading to rapid loss of flavor and color. Blanching also makes the vegetable easier to pack.

Blanchers can be bought, or made from a large kettle [4] and wire mesh holder [3] or basket. There should also be a cover [1]. A wire cover [2] for the basket [3] will help to keep the vegetables down in the water.

All containers and the blanching kettle should be washed and dried in preparation for use.

Sufficient freezer space to store the vegetables should be available.

▶ Preparing Vegetables

Select young tender vegetables for freezer storage. It is better to choose slightly immature vegetables rather than fully ripe. Set aside bruised, damaged, or overripe vegetables for immediate use or discard.

First, wash vegetables thoroughly in cold water.

Then, follow the directions in the accompanying table, which shows how to prepare the individual vegetables.

PREPARATION OF VEGETABLES FOR BLANCHING AND FREEZING

Vegetable	Procedure
Asparagus	Sort by thickness of stalk. Cut off and discard tough parts of stalks. Cut in convenient, equal lengths to fit container.
Beans	
Snap	Cut off ends. Cut in 1- or 2-inch pieces, or slice lengthwise.
Lima	Shell and sort by size, or leave beans in pods to be shelled after blanching.
Beets	Sort by size. Cut off tops, leaving ½-inch stems.
Broccoli	Peel stalks and trim. Split lengthwise into pieces not more than 1 ½ inches across. If necessary to remove insects, soak for 30 minutes in salt water (4 teaspoons salt to 1 gallon of cold water).
Cabbage	Cut off outer leaves. Cut into medium to coarse shreds or thin wedges, or separate head into leaves.
Carrots	Cut off tops and peel. Small carrots may be frozen whole. Cut others into ¼-inch cubes, thin slices, or lengthwise strips.

PREPARATION OF VEGETABLES FOR BLANCHING AND FREEZING (contd)	
Vegetable	**Procedure**
Cauliflower	Break or cut into pieces about ¼-inch across. Peel and split stems. Soak for 30 minutes in salt water (4 teaspoons salt to 1 gallon of cold water).
Celery	Cut stalks into 1-inch lengths.
Chard	Cut off tough stems. Cut leaves into pieces as desired.
Corn	Remove husk and silk. Wash corn.
Eggplant	Peel and cut into ½-inch slices or cubes. Dip in solution of 1 tablespoon lemon juice to 1 quart of water.
Peas	Shell and do not wash.
Peppers, sweet	Cut out stems. Cut peppers in half. Remove seeds and pulp. Cut peppers into ½-inch strips or rings as desired.
Spinach and other Greens (beet, turnip)	Cut off leaves from center stalk. Cut off large leaf stems. Discard injured leaves.
Squash	
Summer	Cut in ½-inch slices.
Winter	Cut into pieces, remove seeds.
Tomatoes	Remove stem ends and cut into quarters. Peel if desired.
Turnips	Cut off tops and peel. Slice into lengthwise strips or chop into small cubes.

▶ Blanching Vegetables

All vegetables except sweet peppers should be "blanched" before freezing. Blanching consists of heating the vegetables in boiling water for a prescribed period of time (blanching time for each vegetable is shown on the next page).

For blanching, a kettle [4] with blanching basket [3], cover [1], and basket cover [2] is used.

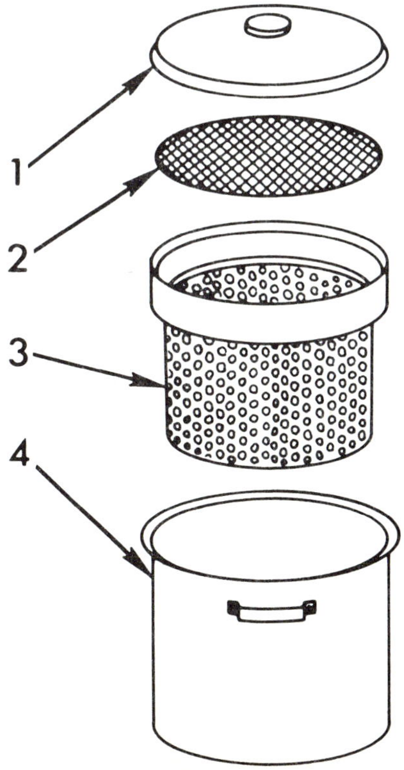

1. For each pound of prepared vegetable, pour 1 gallon or more of water into blanching kettle [4].
2. Bring water to a boil and keep it boiling.
3. Place vegetable in blanching basket [3] and place wire cover [2] on basket. Lower basket into boiling water.
4. Place lid [1] on blanching kettle and start counting time immediately. Keep heat under kettle high for the full blanching time.

The blanching time for each vegetable is shown in the table on the next page.

FREEZING VEGETABLES

Blanching Vegetables

The blanching times are shown in the table.

If you live at 5000 feet or more above sea level, add 1 minute to the blanching time in the table.

For certain vegetables, additional explanation is needed.

Beets, winter squash, and tomatoes are not just blanched, they are thoroughly cooked before freezing, as shown in the table. The special treatment for these vegetables is as follows:

Beets – Cook in boiling water until tender. For small beets, 25–30 minutes; for larger beets, 45–50 minutes.

Winter squash – Cook pieces in boiling water until soft. Remove pulp from rind and mash or press through a sieve into a sauce pan.

Tomatoes – Simmer quarter pieces slowly in a heavy pot, without adding water, until soft, about 20 minutes. Stir continuously to avoid scorching. Seasoning may be added before freezing, but don't add bread crumbs until tomatoes are later heated and served.

BLANCHING TIME FOR VEGETABLES

Vegetable	Blanching Time	Vegetable	Blanching Time
Asparagus	Small stalks – 2 minutes Medium stalks – 3 minutes Large stalks – 4 minutes	Celery	3 minutes
		Chard	2 minutes
Beans Snap	3 minutes	Corn Whole kernel	4 minutes
Lima	Small beans or pods – 2 minutes Medium beans or pods – 3 minutes Large beans or pods – 4 minutes	On-the-cob	Small ears (1-3/4" diameter) – 7 minutes Medium ears (1¼–1½" diameter) – 9 minutes Large ears (1½" and up) – 11 minutes
Beets	Small beets – 30 minutes Medium beets – 45 minutes Large beets – 50 minutes	Eggplant	4 minutes
		Peas	1½ minutes
Broccoli	3 minutes	Peppers	Blanch or not as desired – 2 minutes
Brussels sprouts	Small heads – 3 minutes Medium heads – 4 minutes Large heads – 5 minutes	Spinach and other greens (beet, turnip)	2 minutes
Cabbage	Leaves or shreds – 1½ minutes Wedges – 3 minutes	Squash Summer	3 minutes
		Winter	Cook until soft
Carrots	Small pieces – 2 minutes Large pieces – 3 minutes Whole carrots – 5 minutes	Tomatoes	Cook until soft, about 20 minutes, without water. Stir while cooking
Cauliflower	4 minutes	Turnips	2 minutes

▶ Cooling Vegetables

Immediately after blanching, vegetables should be cooled quickly and thoroughly. This stops the cooking process at the right time.

To cool the vegetables quickly, remove basket [1] of vegetables from boiling water and plunge it into a large quantity of cold water [2] (60°F or below). To keep water cold, use running water or ice water, or change water frequently. It will take about as long to cool the vegetable as it did to heat it.

When the vegetable is cool, remove it from the water and drain thoroughly.

Winter squash and tomatoes are a special case: They were placed in a pan while still hot. Cool the squash or tomatoes by placing the pan containing them in cold water. Stir the squash or tomatoes occasionally.

Except for lima beans, beets, and whole kernel corn, the vegetables are now ready to pack immediately into suitable containers for freezing.

After cooling, lima beans should be removed from the pod, if this was not done earlier.

Beets should be peeled and then cut into slices or cubes.

For whole-kernel corn, now is the time to cut the kernels from the cob.

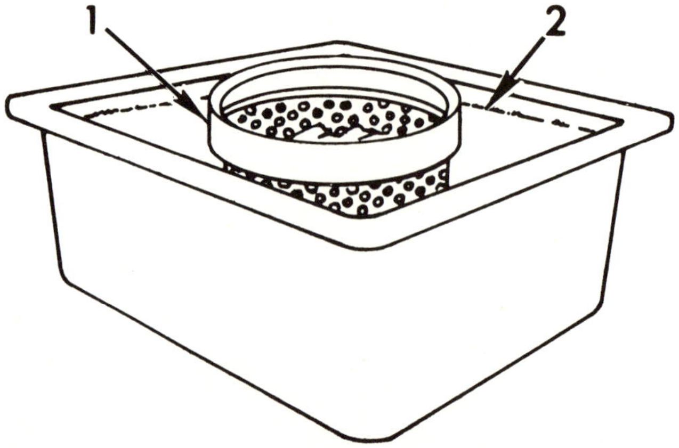

▶ Filling Containers

To assist in packing, a funnel [1] may be used. This will keep the sealing edges of the container clean.

If plastic bags [3] are used as containers, a stand [2] to hold the bag makes filling easier.

When containers are filled, a head space [8] should be left at the top of the container. This is to allow for expansion of the vegetable as it freezes. This head space is not necessary for asparagus, broccoli and corn-on-the-cob, which are packed loosely and need no extra room.

Glass jars [7] require about 1 inch head space [8]. Plastic containers [3] require 1/2 inch head space.

In filling plastic bags [3], carefully work out air pockets. Seal by twisting and folding back top of bag [6], and tying with string [5].

Place labels [4] on all containers. Indicate vegetable and date of freezing.

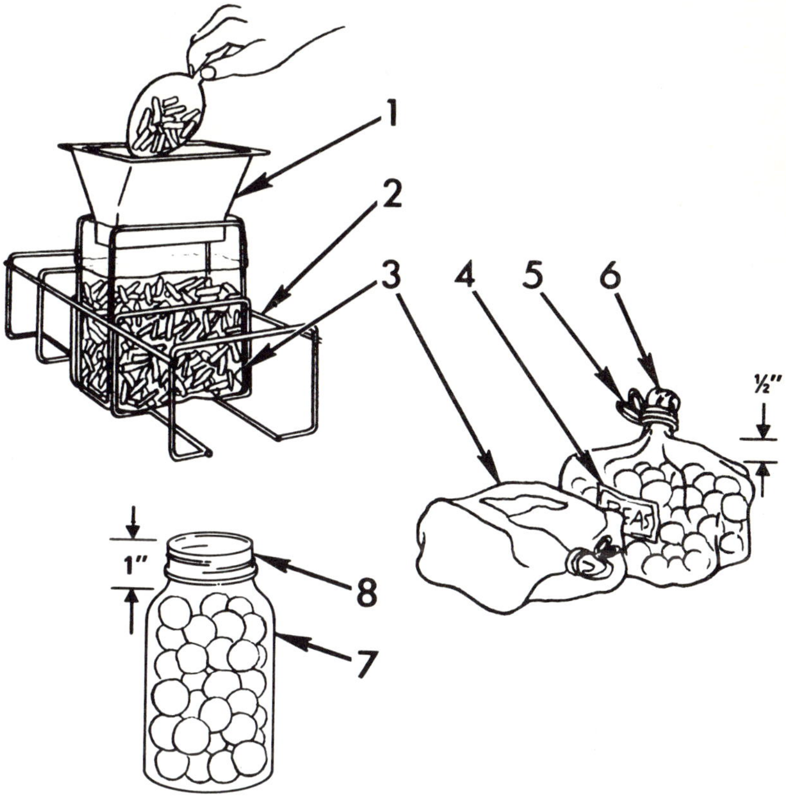

▶ Freezing Vegetables

As soon as the vegetables are packed, they are ready for freezing.

Put no more unfrozen food into a home freezer than will freeze within 24 hours. Usually, this will be about 2 or 3 pounds of vegetables for each cubic foot of freezer capacity. Overloading slows down the rate of freezing, and the vegetables may lose quality or spoil.

If you do not know your freezer's capacity, measure the width (W), height (H) and depth (D) inside the freezer. Measure these dimensions in feet and multiply: W x H x D.

For quickest freezing, leave a little space between packages for air to circulate freely. Packages should be frozen and stored at 0°F or below. At higher temperatures vegetables lose their quality much faster. After they are frozen solid, packages should be stacked closely together.

Most vegetables maintain their high quality for 8 to 12 months at 0°F or below. Longer storage may reduce quality, but will not make vegetables unfit to eat.

It is a good idea to place a list [1] of frozen vegetables near the freezer and keep it up to date. List vegetables as you put them in the freezer, with date. Check vegetables off the list as you remove them.

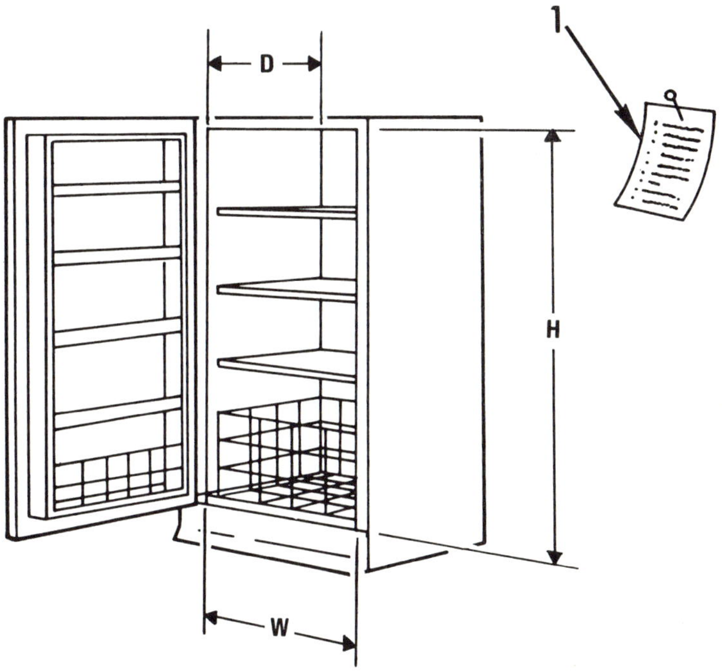

Canning is a preserving process in which food is sealed in air-tight containers and then sterilized by cooking to kill any harmful organism as well as to preserve the condition of the food.

WARNING

Canning directions should be followed carefully to avoid botulism poisoning.

For canning purposes, vegetables fall into two groups which are treated differently:

1. Low-acid vegetables – these include all common vegetables except tomatoes. We will discuss the canning of asparagus, beans, carrots, corn, peas, potatoes, spinach and squash.

2. High-acid vegetables – these include most varieties of tomatoes.

WARNING

To be sure you have high-acid tomatoes, add lemon juice or vinegar to tomatoes as recommended (Page 72).

All common vegetables except tomatoes are low-acid. Because of this, they must be cooked at a temperature (240°F) above that of boiling

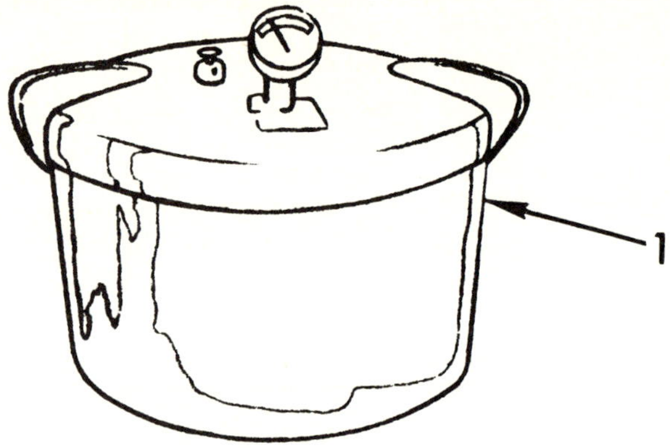

water. This temperature is needed to kill the dangerous botulism organisms. This procedure requires a special item of equipment called a "steam pressure canner" [1].

Tomatoes, being a high-acid vegetable, can be cooked at the temperature of boiling water (212°F). This does not require a steam-pressure canner.

The canning process, once begun, should proceed quickly and smoothly. Vegetables should not be allowed to lie around after harvesting, as they deteriorate. Preferably, vegetables should be canned on the same day they are harvested.

▶ Canning Varieties

Special varieties of vegetables suitable for freezing were listed earlier (Pages 61 and 62). Similarly, certain special varieties are suitable for canning. These varieties are shown in the accompanying table. Some varieties appear on both this list and the previous list for freezing.

The seed companies from which the individual varieties can be ordered are shown on the right hand side of the table. The X's in the table show which companies handle which varieties.

The addresses of the seed companies are given on Page 61.

GOOD VEGETABLE VARIETIES FOR CANNING

Vegetable	Varieties	Burgess	Burpee	Farmer	Field	Gurney	Harris	Jung	May	Shumway
Beans										
Snap										
Bush	Blue Lake				X	X		X	X	
	Brittle Wax			X	X			X		
	Butter Wax					X	X		X	
	Contender					X			X	
	Garden Green					X				
	Green Isle			X						
	Improved Tendergreen			X	X	X			X	X
	Kinghorn Wax		X			X	X			
	Leka Lake		X							
	Tenderette				X	X	X			X
	Top Crop	X	X	X		X			X	
Pole	Blue Lake				X			X	X	
	New Pole			X						
	Purple Pod				X					
Lima	Baby Bush		X		X			X		
	Burpee's Best		X							
	Clark's Green Seeded					X				
	Fordhook 242	X	X		X	X	X	X	X	X
	Henderson's		X		X				X	X

Canning Varieties

GOOD VEGETABLE VARIETIES FOR CANNING (contd)

Vegetable	Varieties	Burgess	Burpee	Farmer	Field	Gurney	Harris	Jung	May	Shumway
Beets	Early Wonder			X	X	X		X	X	X
	Pacemaker					X	X			
	Red Ball		X							
	Red Ruby Queen					X			X	
	Sweetheart			X						
Carrots	Gold Pak		X				X			
	Nantes Half Long		X				X			
Corn	Country Gentleman Hybrid		X		X	X			X	X
	Early Golden Maincrop				X					
	Golden Beauty			X						
	Golden Cross Bantam		X	X	X	X				X
	Iochief		X	X	X	X	X		X	X
	Jubilee			X						
	Kanner King			X						
	Marcross Hybrid					X				
	Most Tender				X					
	New Cheddar Cross Hybrid							X		
	Stowell's Evergreen		X	X		X			X	X
	White Evergreen									X
Peas	Alaska	X				X				X
	Early Allsweet					X				
	Feeezonian	X	X	X	X		X			X
	Giant Stride	X								
	Laxton's Progress	X		X	X	X				X
	Little Marvel	X	X	X		X			X	
	Victory Freezer					X				
	Wando	X	X	X	X		X	X		
Peppers										
Sweet	Sunnybrook									X
	Worldbeater		X							
Hot	Hungarian Wax	X	X		X	X				X
	Pimiento	X			X				X	X
Spinach	Giant Thick Leaf				X					
	Hybrid No. 7		X	X						
	New Hybrid					X				
	Northland					X				
	Virginia Blight Resistant		X							
Squash										
Winter	Golden Delicious					X	X			
	Hubbard (Golden or Red)	X	X	X						
Tomatoes	Beefsteak					X			X	
	Colossal	X								
	Crimson Giant					X				
	Garden State									X
	Italian Canner		X							X
	Marglobe	X	X		X					X
	Pink Gourmet								X	
	Pinkshipper	X								
	Ponderosa			X		X			X	
	Red Pear				X					X
	Red Sugar	X								
	Roma		X				X		X	X
	Rutgers	X	X	X	X		X		X	X
	Yellow Pear								X	

CANNING VEGETABLES

▶ Preparing Equipment

All equipment should be clean and available before you begin the canning process.

Containers

The most common containers for home canning are glass canning jars. Glass canning jars may be standard [1] or wide-mouth [4].

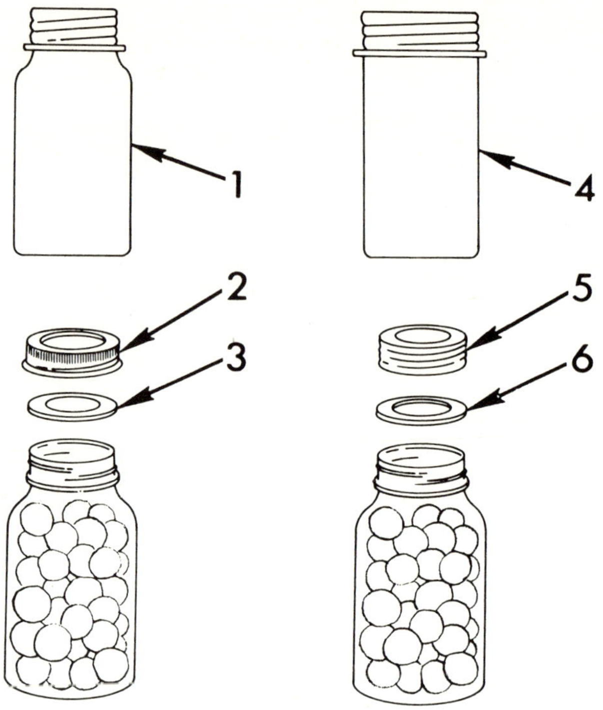

WARNING

Be sure to use special canning jars for canning. Do not use ordinary glass jars. Food may spoil.

Lids (or "closures") for glass canning jars are of two main types:

The first type is a metal screwband [2] and flat metal lid [3]. A sealing compound is used.

The second type is a porcelain-lined screw cap [5] with a shoulder rubber ring [6].

Metal screwbands [2] and porcelain-lined screw caps [5] can be reused if in good condition. Metal lids [3] and rubber rings [6] should be used only once.

Preparing Equipment

Check old containers before use. Discard any with cracks, chips, dents, or rust.

To assist you in determining your container requirements, the accompanying table shows the approximate amount of fresh vegetables that make 1 quart of canned vegetable.

Amount of Fresh Vegetable to Make 1 Quart of Canned Vegetable

Vegetable	Pounds
Asparagus	2½–4½
Beans	
Snap	1½–2½
Lima, in pods	3–3½
Beets, without tops	2–3½
Carrots, without tops	2–3
Corn, in husks	3–6
Peas, in pods	3–6
Spinach	2–6
Squash	
Summer	2–4
Winter	1½–3
Tomatoes	2½–3½

Steam-Pressure Canner

The steam-pressure canner (4) should be clean and ready. Clean the petcock (1) by drawing a string through the openings (2).

Check the pressure gage (3). The gage may be a dial-type (3) or a weighted type. A dial-type should be checked by your county home demonstration agent, dealer, or manufacturer. A weighted type should be thoroughly clean.

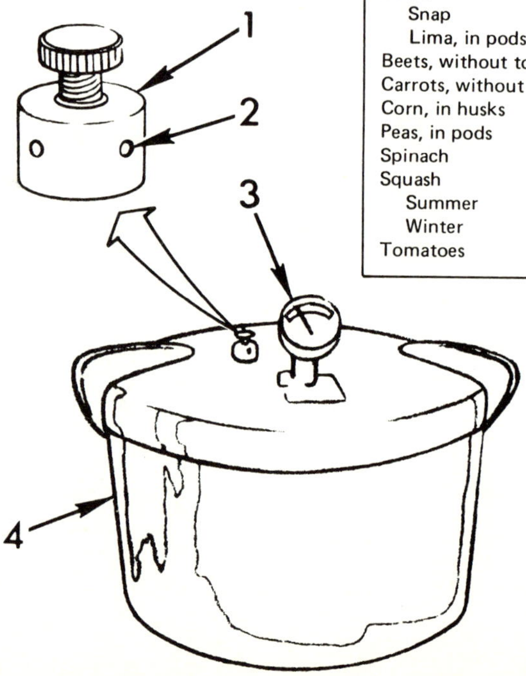

Preparing Equipment

Wash canner kettle [1] thoroughly. Wipe lid [2] with soapy cloth, then with damp, clean cloth. Do not place in water. Dry kettle and lid with dry cloth.

Water-Bath Canner

For canning tomatoes, a water-bath canner [4] is used instead of a steam-pressure canner. Any large metal container can be used for this purpose provided it is deep enough. Also, you will need a cover [3] and a rack [5].

To check depth, allow 1 inch for rack [6] at bottom, then 5 to 7 inches for container [7], then 2 to 4 inches space above container. This space is needed for water [8] above the container and for boiling space [9]. Total depth should be at least 10 to 12 inches, depending on container.

If your steam-pressure canner is deep enough, you can use it as a water-bath canner. In this use the cover is not fastened and the petcock is left wide open.

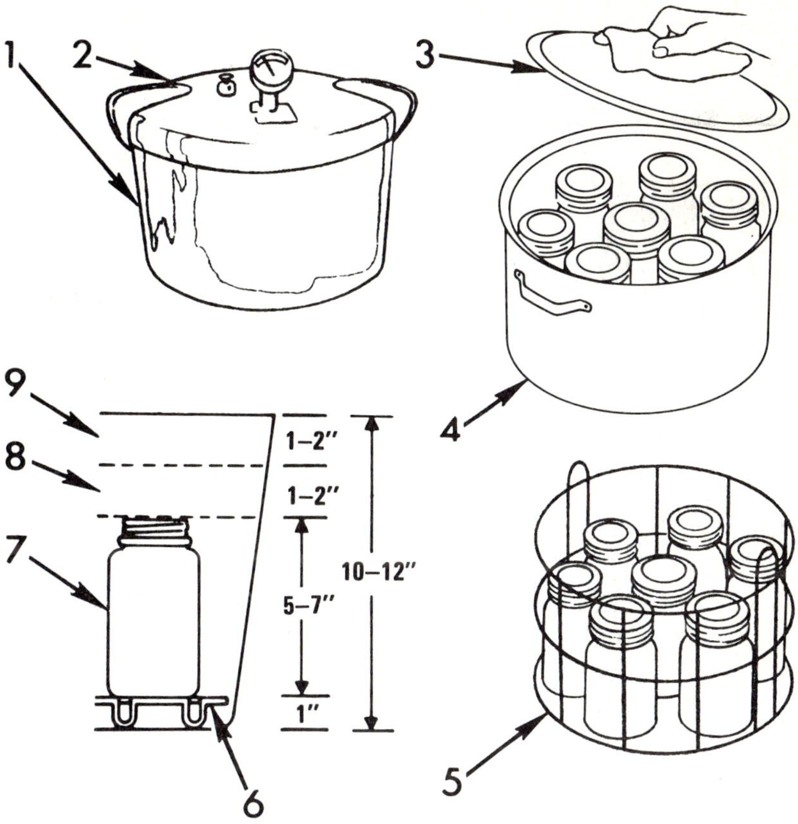

▶ Preparing Vegetables

Select only the best vegetables in good condition. There should be no damage or spoiled parts.

First, wash vegetables thoroughly in cold water. Sort for size and maturity. This will help uniform heating and easy packing.

Then follow the directions in the accompanying table, which shows special preparation procedures for each vegetable.

PREPARATION OF VEGETABLES FOR CANNING

Vegetable	Procedure
Asparagus	Trim off scales and tough ends and wash again. Cut into 1 inch pieces.
Beans	
Snap	Trim ends. Cut into 1 inch pieces.
Lima	Shell and wash.
Beets	Cut off tops leaving an inch of stem. Also leave root.
Carrots	Scrape, then slice, dice, or leave whole.
Corn	Remove husk and silk. Wash. Cut kernels from cob.
Peas	Shell and wash.
Potatoes	Peel. Cut into ½-inch cubes or leave whole up to 2½ inches diameter (as desired).
Spinach	Cut out tough stems.
Squash	
Summer	Trim ends. Cut into ½-inch slices, then cut into halves or quarters to make pieces of uniform size.
Winter	Cut in half and remove seeds. Remove rind. Cut into 1 inch cubes.
Tomatoes	To loosen skins, dip into boiling water for ½ minute. Then dip into cold water. Cut out stem ends and peel tomatoes.

CANNING VEGETABLES

▶ Filling Containers

Using hot, soapy water, wash glass jars and closures. Rinse well. Place in hot water until needed.

There are two methods that may be used for filling the containers, the raw pack method and the hot pack method. In the raw pack method (also called the cold pack method), the raw vegetable is placed directly in the jars. In the hot pack method, the vegetable is first cooked for a short time before being placed in the jars.

Either method can be used. With the hot pack method, somewhat more vegetables can be packed in each container. However, the additional heating may result in some loss of food value.

If using the hot pack method, see section below.

Raw Pack Method

Place the prepared vegetable (see table, Page 71) in the jar. The accompanying table shows the amount of space to leave for each vegetable. If desired, add 1/2 teaspoon salt to pint jar or 1 teaspoon salt to quart jar.

After packing the vegetable, pour boiling water into the jar until it just covers the vegetable. If the water does not cover the vegetable, the vegetable will become discolored.

WARNING

Some varieties of tomatoes are low-acid. Adding lemon juice or vinegar will increase the acid content, reducing the danger of botulism.

Add lemon juice or vinegar to each jar of tomatoes:

2 teaspoons to a pint
1 tablespoon to a quart

The next step is to install the closures. Go to Page 73.

PACKING SPACE FOR RAW PACK METHOD

Vegetable	Packing Space
Asparagus	Pack tightly. Leave ½ inch at top.
Beans	
Snap	Pack tightly. Leave ½ inch at top.
Lima	Pack loosely. Leave 1 inch at top of jar for pints and 1-½ inches for quarts.
Carrots	Pack tightly. Leave 1 inch at top.
Corn	Pack loosely. Leave 1 inch at top.
Peas	Pack loosely. Leave 1 inch at top.
Squash	
Summer	Pack tightly. Leave 1 inch at top.
Tomatoes	Leave tomatoes whole or cut into halves or quarters. Pack loosely. Leave ½ inch at top.

Filling Containers

Hot Pack Method

In this method the prepared vegetable (see table, Page 71) is first cooked a certain amount of time before being placed in the containers. The accompanying table shows how to cook each vegetable.

After cooking, place the vegetable in the jars.

Vegetables being hot-packed should always be packed loosely. The amount of space to leave at the top of the jar is as follows: For lima beans, corn, and peas, leave 1 inch. For all other vegetables, leave 1/2 inch.

As with the raw pack method, add salt if desired. Add 1/2 teaspoon salt to a pint jar and 1 teaspoon salt to a quart jar.

After packing the vegetable in the jar, pour boiling water into the jar until it just covers the vegetable. Use the water in which the vegetable was cooked for this purpose. If there is not enough, then add boiling water as needed.

WARNING

Some varieties of tomatoes are low-acid. Adding lemon juice or vinegar will increase the acid content, reducing the danger of botulism.

Add lemon juice or vinegar to each jar of tomatoes:

2 teaspoons to a pint
1 tablespoon to a quart

COOKING PROCEDURES FOR HOT PACK METHOD

Vegetable	Procedure
Asparagus	Cover with boiling water. Boil 2 to 3 minutes.
Beans	
Snap	Cover with boiling water. Boil 5 minutes.
Lima	Cover with boiling water, bring to boil.
Beets	Cover with boiling water. Boil 15 to 25 minutes, until skin slips easily. Remove skin. Trim off stem and root. Leave baby beets whole. Cut medium or large beets in ½ inch cubes or slices.
Carrots	Cover with boiling water, bring to boil.
Corn	To each quart of corn add 1 pint boiling water. Bring to boil.
Peas	Cover with boiling water, bring to boil.
Potatoes	
Cubed	Dip cubes in brine (1 teaspoon salt to 1 quart water) to prevent darkening. Drain. Cover with boiling water. Boil 2 minutes.
Whole	Cover with boiling water. Boil 10 minutes.
Spinach	Place spinach in a cheesecloth bag and steam for 10 minutes or until well wilted.
Squash	
Summer	Cover with boiling water, bring to boil.
Winter	Cover with boiling water, bring to boil.
Tomatoes	Cut into quarters and bring to boil. Stir to keep tomatoes from sticking.

▶ **Installing Closures**

If using porcelain-lined screw cap, see section below.

To install metal screwband and flat metal lid with sealing compound, after vegetable is packed:

1. Wipe jar rim [3] clean.
2. Apply sealing compound to lid [2].
3. Place lid [2] on jar [4], with sealing compound underneath, next to jar rim [3].

When band is tight, this lid has enough "give" to let air escape during processing.

4. Screw metal band [1] down tight by hand.

If canning tomatoes, go to Page 75 for processing with water-bath canner.

If canning any vegetable but tomatoes, go to Page 74 for processing with steam-pressure canner.

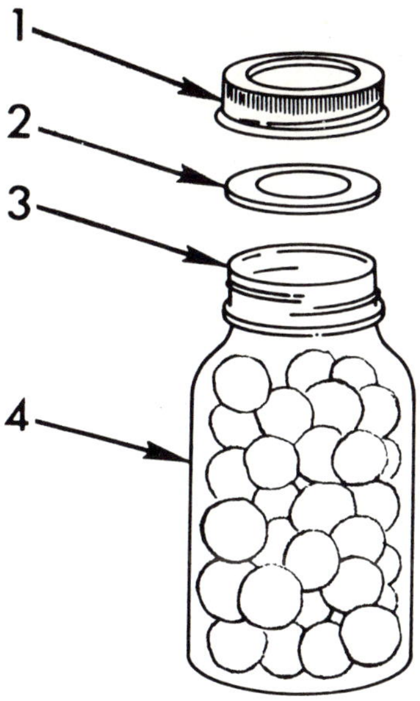

Installing Closures

To install porcelain-lined screw cap:

1. Wipe jar rim [3] and shoulder [4] clean.
2. Moisten rubber ring [2] and place it on jar shoulder [4].

Cap [1] will be tightened again later, after processing.

3. Screw cap [1] down firmly, then turn back 1/4 inch.

If canning tomatoes, go to Page 75 for processing with water-bath canner.

If canning any vegetable but tomatoes, go to next page for processing with steam-pressure canner.

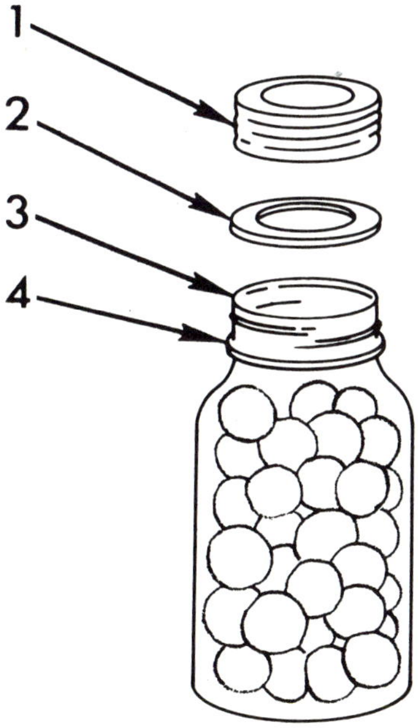

CANNING VEGETABLES

▶ Processing with Steam-Pressure Canner

The following procedure is used for all vegetables but tomatoes. For tomatoes, use the procedure on Page 75, Processing with Water-Bath Canner. In general, follow the manufacturer's directions for the steam-pressure canner you are using.

WARNING

Working with steam pressure can be dangerous. Be sure to follow directions carefully.

1. Put 2 to 3 inches of water in the bottom of canner [1]. Bring to boil.

In placing jars in canner, place them in positions so that steam can flow around each jar.

2. Place filled glass jars [2] on rack in canner. If more than one layer of jars is used, use another rack.

3. Fasten canner cover [3] securely. No steam should escape except through vent [4] (petcock or weighted-gage opening).

4. When steam comes steadily from vent, allow it to continue for 10 minutes.

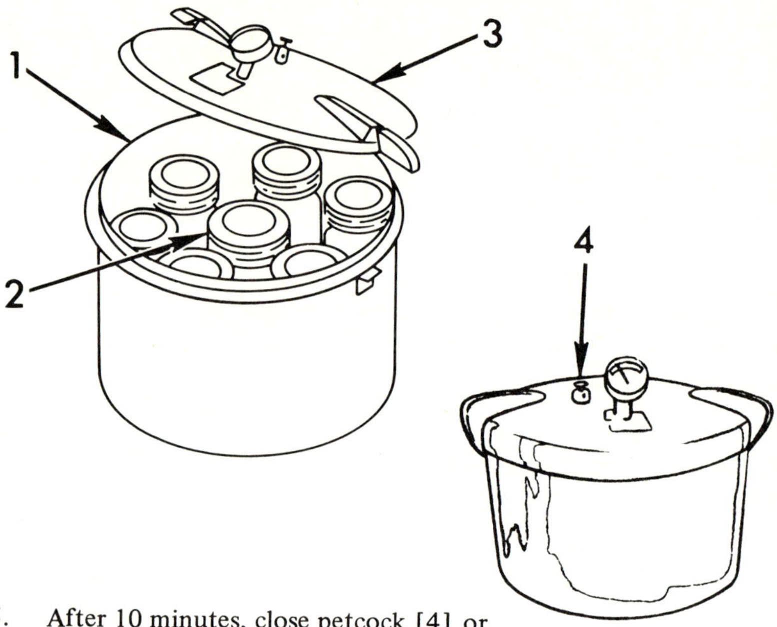

5. After 10 minutes, close petcock [4] or install weighted gage.

6. When pressure rises to 10 pounds (240°F), start counting processing time (see Page 75 for processing time).

Processing with Steam-Pressure Canner

7. Keep pressure constant by adjusting heat under canner. Do not lower pressure by opening petcock [1].

8. When processing time is up, remove canner from heat immediately (see Page 75 for processing time).

WARNING

Steam pressure is dangerous. Follow Steps 9, 10, and 11 carefully.

9. Let canner stand until pressure is zero.

Do not pour cold water over canner.

10. When pressure is zero, wait one or two minutes. Then slowly open petcock [1] or remove weighted gage.

11. Unfasten cover and tilt far side [2] up so steam escapes away from you. Set cover aside.

WARNING

Jars are hot. Use tongs or potholder for following two steps.

Do not place jars on cold surface. Leave space between jars so air can circulate.

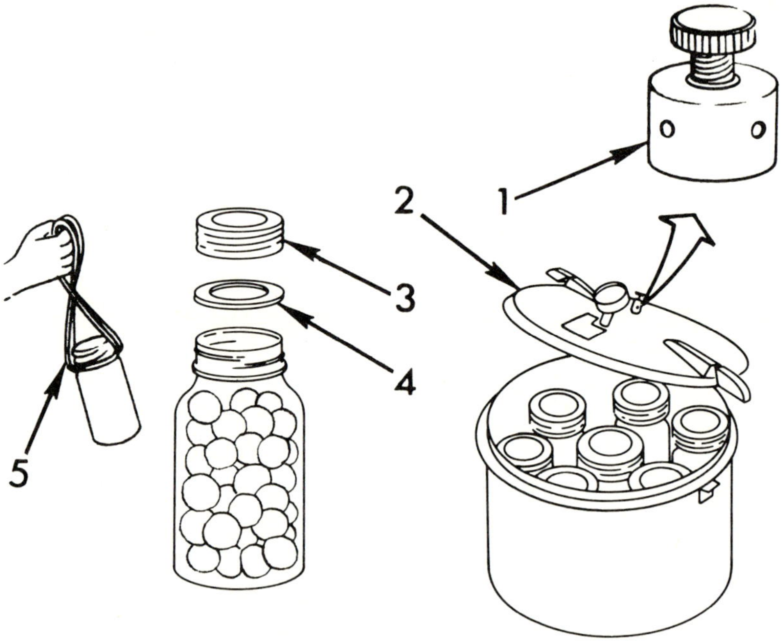

12. Using tongs [5] or potholder, remove jars from canner and place on tray, rack, or folded cloth.

13. If jars have porcelain–lined screw cap [3] with rubber ring [4], use potholder and screw caps [3] down tight to complete seal.

Go to Cooling Canned Vegetables, Page 76.

Processing with Steam-Pressure Canner

Processing Times

The processing times for the steam-pressure canner are given in the accompanying table. They apply to any altitude. However, if you live at higher altitudes you will need to process your vegetables at a higher pressure than 10 pounds. For each 2000 feet altitude, add 1 pound of pressure:

Altitude (ft)	Processing Pressure (lbs)
0–2000	10
2000–4000	11
4000–6000	12
6000–8000	13
8000–10,000	14
10,000 and above	15

PROCESSING TIMES IN STEAM-PRESSURE CANNER

Vegetable	Time (in minutes) Pint Jar	Quart Jar
Asparagus	25	30
Beans		
Snap	20	25
Lima	40	50
Beets	30	35
Carrots	25	30
Corn	55	85*
Peas	40	40
Potatoes		
Cubed	35	40
Whole	30	40
Spinach	70	90
Squash		
Summer – Raw Pack**	25	30
– Hot Pack**	30	40
Winter	55	90

*The long processing time for corn in quart jars leads to discoloration. Therefore, pint jars are recommended.

**Summer squash is the only vegetable for which the processing time differs for raw pack and hot pack vegetables.

▶ Processing with Water-Bath Canner

The following procedure is used only for tomatoes. For other vegetables, use the procedure on Page 74, Processing with Steam-Pressure Canner.

If canning jars were filled with raw pack tomatoes, go to Step 2. If canning jars were filled with hot pack tomatoes, start with Step 1.

1. Fill water-bath canner [3] about half full of water. Bring to boil. Go to Step 3.
2. Fill water-bath canner [3] about half full of hot, but not boiling, water.
3. In a separate kettle, start bringing extra water to a boil. It may be needed later (Steps 5 and 7).
4. Place filled jars [2] on rack in canner [3]. If more than one layer of jars is used, use another rack.

CAUTION

Do not pour boiling water directly on jars.

5. Add boiling water if needed to bring water level 1 or 2 inches above top of jars [2]. Place cover [1] on canner.
6. When water in canner comes to a boil, start to count processing time (See section above for processing time).

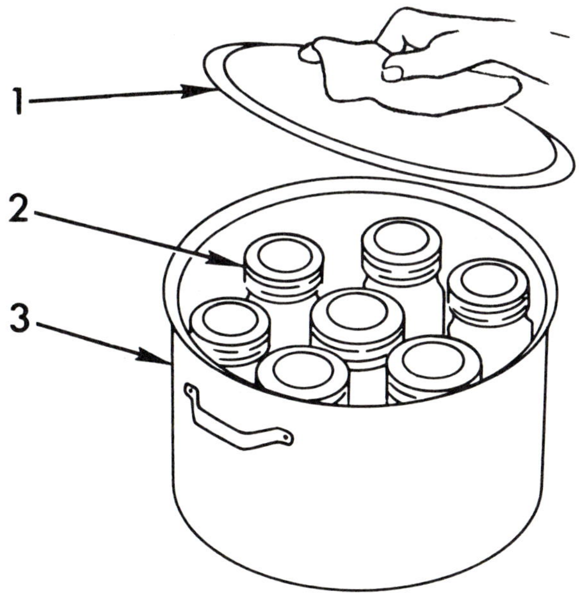

CANNING VEGETABLES

Processing with Water-Bath Canner

7. Add boiling water during processing if needed to keep containers covered.

WARNING

Jars are hot. Use tongs or potholder for following two steps.

Do not place jars on cold surface. Leave space between jars so air can circulate.

8. When processsing time is up, use tongs [1] or potholder to remove jars [2] from canner [3] and place jars on tray, rack, or folded cloth. (See table for processing time)

9. If jars have porcelain-lined screw cap [5] with rubber ring [4], use potholder and screw cap [5] down tight to complete seal.

Processing Times

For tomatoes, the processing times are as follows:

Raw Pack	–	pint jars, 35 minutes
		quart jars, 45 minutes
Hot Pack	–	pint jars, 10 minutes
		quart jars, 10 minutes

These times apply to processing of tomatoes at sea level. For each 1000 feet of altitude, add 2 minutes to the processing time for raw pack tomatoes and 1 minute to the processing time for hot pack tomatoes:

	Processing Times (minutes)		
	Raw Pack		Hot Pack
Altitude (ft)	Pints	Quarts	Pints or Quarts
0–1000	35	45	10
1000–2000	37	47	11
2000–3000	39	49	12
3000–4000	41	51	13
4000–5000	43	53	14
5000–6000	45	55	15
6000–7000	47	57	16
7000–8000	49	59	17
8000–9000	51	61	18
9000–10,000	53	63	19

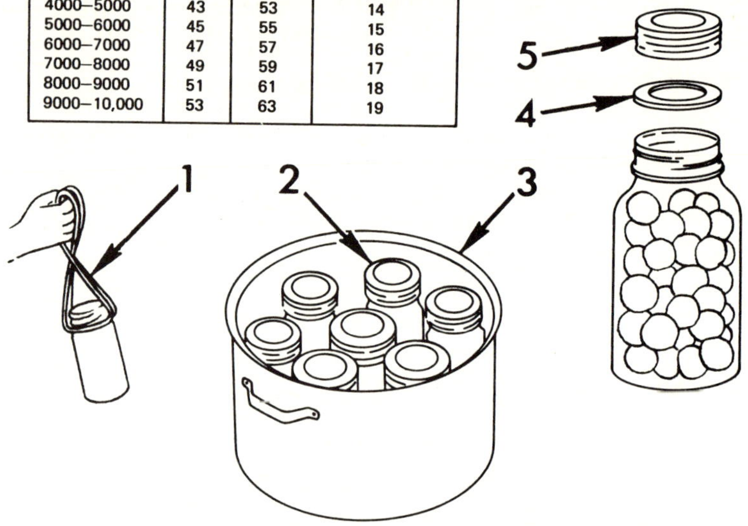

▶ Cooling Canned Vegetables

As jars are removed from the canner (steam-pressure or water-bath), the porcelain-lined screw caps must be tightened, as described. If liquid boiled out during processing, do not open jar to add more.

Do not place hot jars [1] on cold surface. Place them on a tray, rack [2], or folded cloth.

Jars should be cooled top side up. Allow room around each jar for air to circulate.

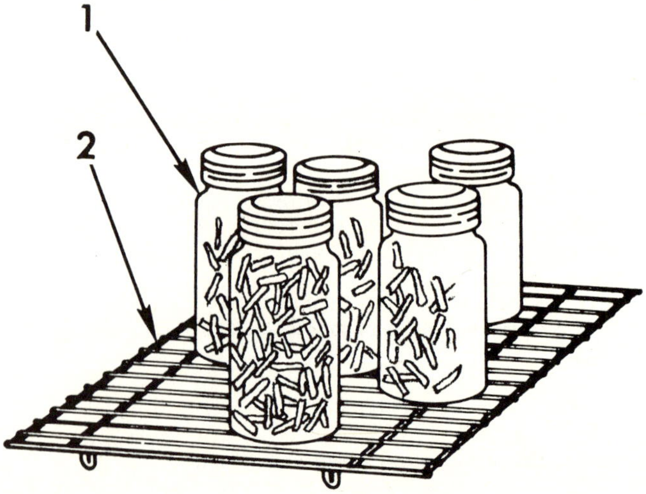

▶ Checking and Labeling Before Storing

The day after canning, check the seals on the jars.

To check a jar that has a metal screwband [1] and flat metal lid [2], press center of lid [2]. If lid does not move, jar is sealed. Carefully remove screwband [1], wash and store in a dry place.

To check a jar that has a porcelain-lined screw cap [4] and rubber ring [5], turn each jar partly over in your hands. Check for leak. If no leak, jar is sealed.

Before storing jars, wipe outside of jars clean. Apply labels [3,6] to show contents, date, and "batch number". The batch number is useful if you made more than one batch in a day, and if there is later any spoilage problem.

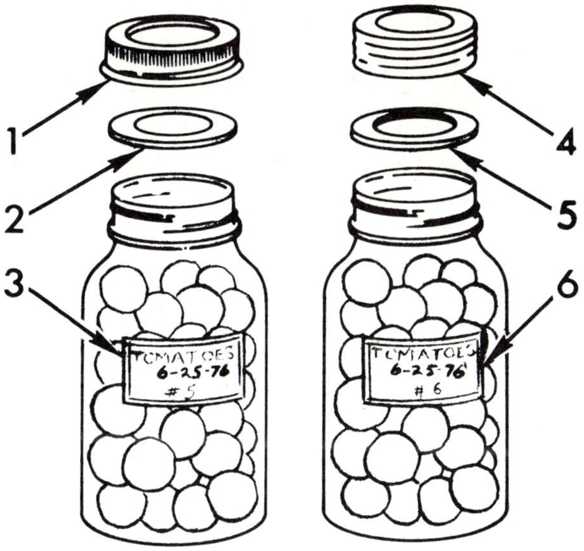

▶ Storing Canned Vegetables

Properly canned vegetables will remain good to eat for a year if stored in a cool, dry place. Canned vegetables will lose some of their eating quality in a few weeks or months, if stored in a warm place near a stove, furnace, hot pipes, or in direct sunlight.

Dampness may cause corrosion of metal lids. The food may leak and spoil.

If canned vegetables are frozen, they will not spoil unless the seal is damaged or the jar is broken. However, the eating quality will not be as good. In an unheated storage space, the possibility of freezing can be reduced by wrapping jars in paper or covering them with a blanket.

▶ Checking Canned Vegetables Before Use

Before opening each jar, look at it carefully. Check for bulging lids or rings, or a leak. These may mean the seal has broken and the vegetable has spoiled.

When you open the jar, look for other signs—spurting liquid, an off odor, or mold.

WARNING

If you suspect that the vegetable has spoiled, do not test it by tasting it. Botulism bacteria are so poisonous that a taste can be fatal.

Boil the suspected vegetable for a few minutes. If you notice an unusual or unpleasant odor, the vegetable is not safe to eat. Burn it or dispose of it so that it will not be eaten by any person or animal.

Drying is a less-common method of preserving for vegetables than it is for fruit. However, as one of the oldest methods of preservation, it has regained popularity. Beans, peas, and particularly herbs are good for drying.

Drying is a simple process in which water is removed and nothing is added. As in freezing, vegetables (except onions) are blanched before drying. This stops the growth processes at the right time and prevents undesirable changes during storage. It also shortens the drying time by softening the tissues. Because the blanching time is longer for drying than for freezing, steam rather than boiling water is used. This reduces the loss of nutrients.

Drying can be done outside on trays [1] or inside in an oven. Also, special dehydrators [3] can be bought.

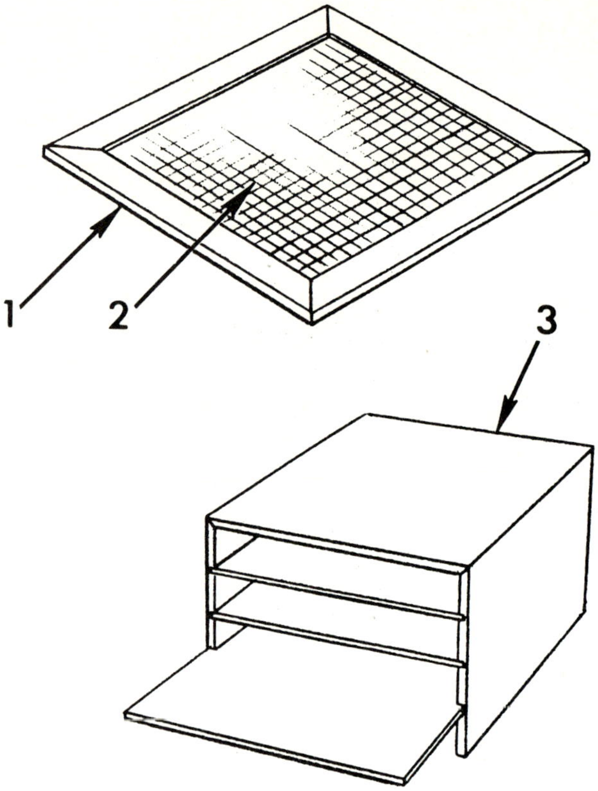

▶ Preparing Equipment

Tray: Vegetable-drying trays can be either bought or made. They are simple to make from wood [1] and screen [2], wire mesh, or cheese cloth.

Preparing Equipment

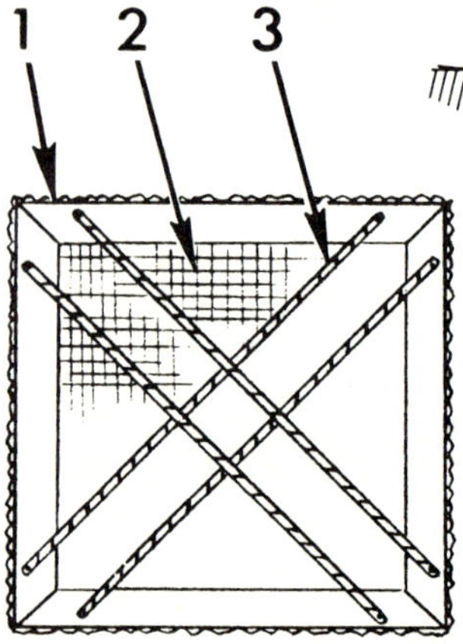

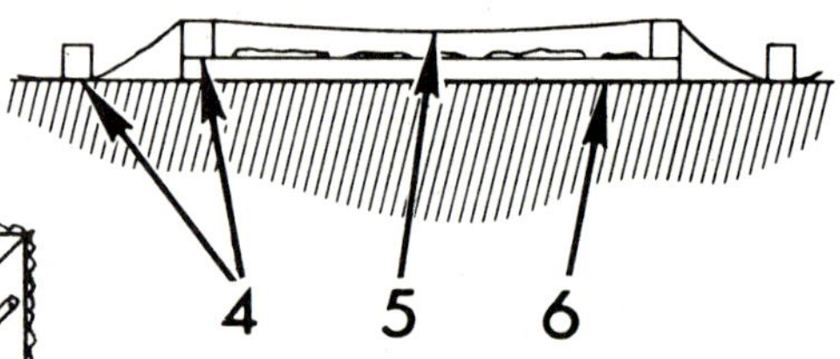

A tray can be made by joining 4 pieces of wood [1] into a frame as shown. Screen [2], wire mesh, or cheesecloth is then stretched over the frame and tacked down. For support, string [3] or wire is stretched from corner to corner.

If cheesecloth is used to make the tray, the frame should be made about 34 inches square. This is because cheesecloth normally comes in 36-inch bolts.

The screen, wire mesh, or cheesecloth allows air to circulate freely and dry the vegetables.

Another piece of cheesecloth [5] is then placed over the tray [6], and supported by blocks [4]. This keeps insects and dirt off the tray.

Clean window screens [7] can also be used as drying trays. Two screens, one on top of the other, separated by blocks of wood [8], make a good dry tray.

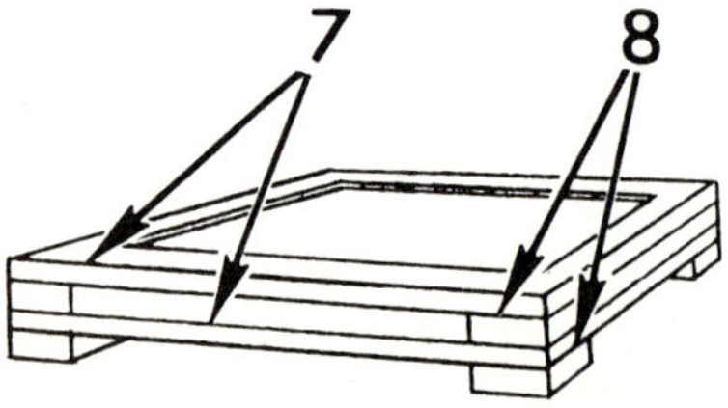

If drying is done in an oven or dehydrator, special trays are not needed.

Blanching Kettle: For blanching, a large pot [11] with wire rack [10] and lid [9] will be needed.

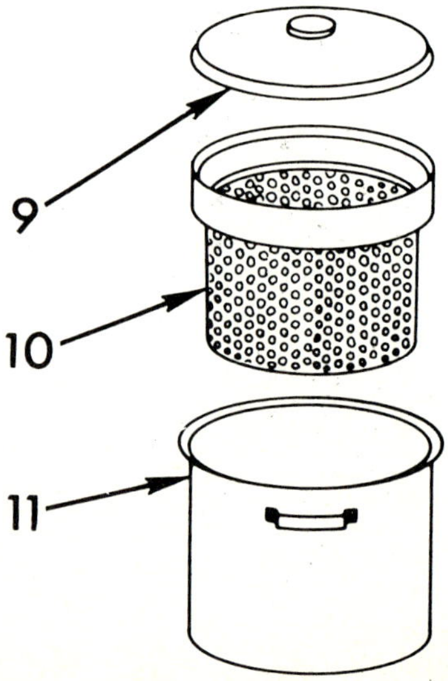

▶ Preparing Vegetables

First, wash vegetables thoroughly in cold water.

Then, follow the directions in the accompanying table for preparing vegetables for blanching and drying.

Herbs are discussed later (Page 81).

PREPARATION OF VEGETABLES FOR BLANCHING AND DRYING

Vegetable	Procedure
Asparagus	Cut off and use only top 3 inches of spear.
Beans	
Snap	Cut or slice into small pieces.
Lima	Remove beans from pods, discard pods.
Beets	Remove tops and roots.
Broccoli	Slice into ½-inch strips.
Brussels sprouts	Cut into ½-inch slices.
Cabbage	Cut into long thin slices.
Carrots	Cut into slices or quarters.
Celery	Remove leaves. Cut stalks into small pieces.
Chard	Cut into wide strips.
Corn	Remove husk and silk.
Onions, bulb	Peel and cut into small strips.
Peas	Remove peas from pods, discard pods.
Peppers	Cut into thin strips.
Spinach	Cut into wide strips.
Squash	
Summer	Cut into thin strips.
Winter	Cut open, remove pulp and seeds. Cut into 1-inch strips. Then peel.

▶ Blanching Vegetables

To blanch vegetables in preparation for drying, proceed as follows:

1. Pour 2 to 3 inches of water in pot [1] and bring to boil. Water level [3] should be below basket [2].

2. Place shallow layer of vegetables about 2 inches deep in basket.

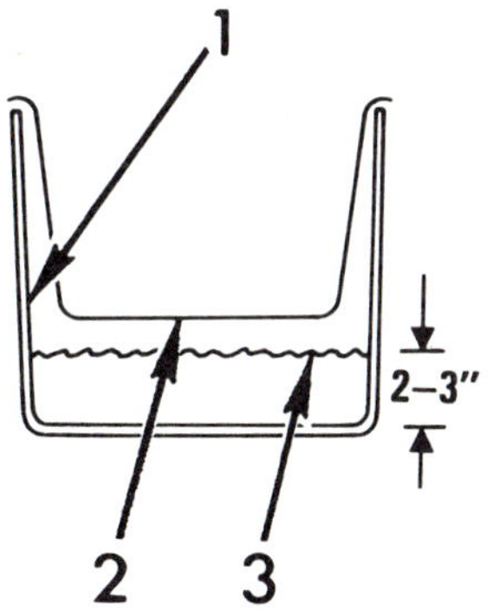

Blanching times for individual vegetables are shown in accompanying table. No adjustment in time is needed if you live at a high altitude.

3. Place cover on pot and boil rapidly for full blanching time.

Vegetables are now ready for drying except for beets and corn:

beets – cool, peel and cut into small cubes or thin slices

corn – cool, then cut kernels off of cob

BLANCHING TIME FOR VEGETABLES

Vegetable	Blanching Time (in minutes)
Asparagus	10
Beans	
Snap	20
Lima	20
Beets	45 or until cooked through
Broccoli	10
Brussels sprouts	12
Cabbage	8
Carrots	10
Celery	10
Chard	5
Corn	10
Onions, bulb	8*
Peas	15
Peppers	10
Spinach	5
Squash	
Summer	7
Winter	10

*Blanch onions only if they are to be cooked before eaten.

DRYING VEGETABLES

▶ Drying Vegetables

Usually, the vegetable is placed one layer thick on the trays. For leafy vegetables like cabbage, chard, and spinach, the layer should not be more than 1/2-inch thick.

1. Place pieces of vegetable [1] on tray [2].

2. Place tray in dry, protected area, such as screen porch.

3. Place piece of cheesecloth [3] or similar mesh over, but not touching vegetables on tray [5]. Use bricks [4] or wooden blocks for support.

The top piece of cheesecloth will protect the vegetables from insects, but not from moisture. The trays may be placed outdoors on sunny days, but should be brought indoors at night.

Not all pieces of vegetable will finish drying at the same time. Vegetables are through drying when they are brittle.

4. Remove pieces of vegetable as they dry.

If drying is done in an oven, set the oven temperature as low as possible. Usually, this will be about 200°F. An advantage of a special dehydrator [6] is that it can be set at a lower temperature. 110°–120°F is best.

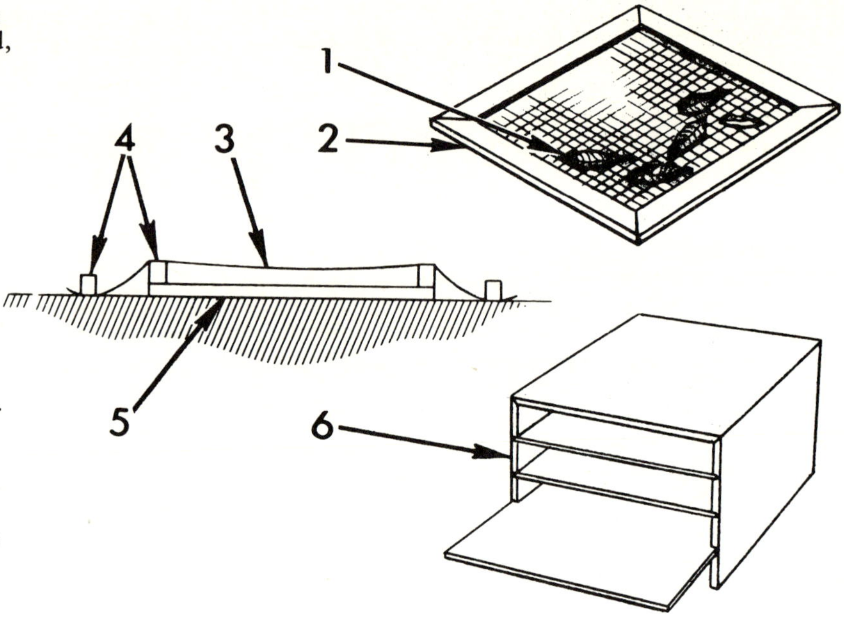

Using an oven or dehydrator, drying can take place in all kinds of weather, and does not take as long.

▶ Storing and Rehydrating Vegetables

When finished drying, store the vegetables in plastic bags [2] or canning jars [1] with good lids.

If plastic bags are used, close them tightly with wire, string [3], or rubber bands.

Store the containers in a cool dry area. In warm, humid weather, store the containers in the refrigerator.

Dried vegetables can be eaten raw, but you will usually want to add the water that was removed in drying. That is, you will want to rehydrate.

To rehydrate vegetables, pour 1-1/2 cups of boiling water over each cup of dried vegetable. Let stand until all the water is absorbed. This will usually take about 2 hours, but depends on the size of pieces and their dryness. If the water is absorbed quickly, add water slowly until no more is absorbed.

The vegetables are now ready for cooking in the normal manner: add water, bring to boil, cook until tender.

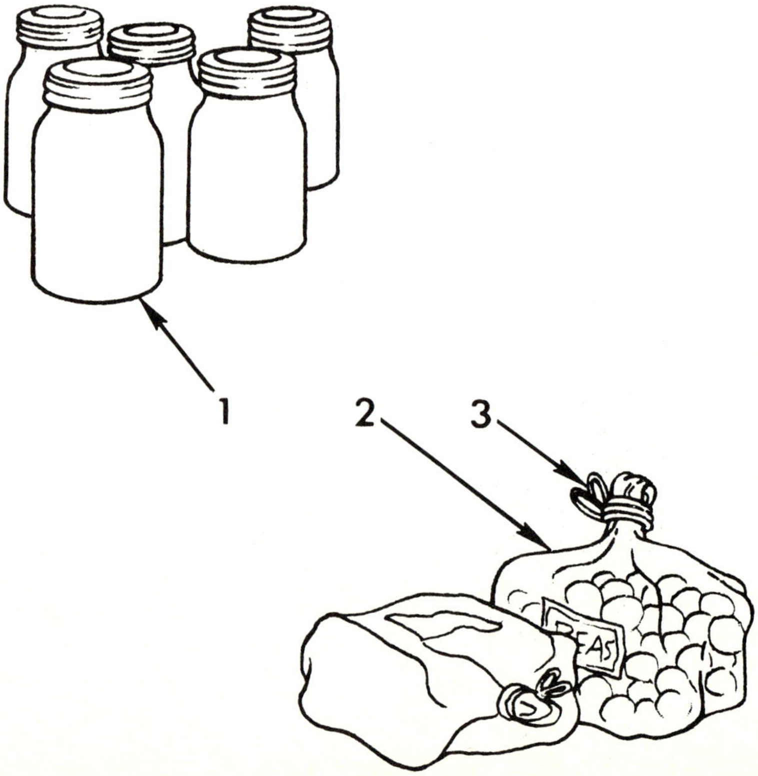

▶ Drying Herbs

Herbs can be dried either by means of a tray, as with vegetables, or by hanging them in a dry place.

If you are going to hang herbs instead of using the tray, see section below.

To use the tray, place the herbs [1] on the tray [2] one layer deep. Put a separate piece of cheesecloth [3] or similar material over, but not touching, the herbs. Blocks of wood [4] or stones can be used for this purpose.

The trays can be placed outdoors on a sunny day, but should be brought inside overnight. Indoor drying is better. A screen porch is an ideal spot to place the tray and leave it.

In the tray, herb leaves will dry quickly, in 3 or 4 days, if conditions are right. Then the flavor is good. If the leaves are not dry in two weeks, place them in an oven at 100°F. Keep them there until they crumble into dust between the fingers. Then take them out and remove the leaves from the stems.

Go to Page 82 for storing herbs.

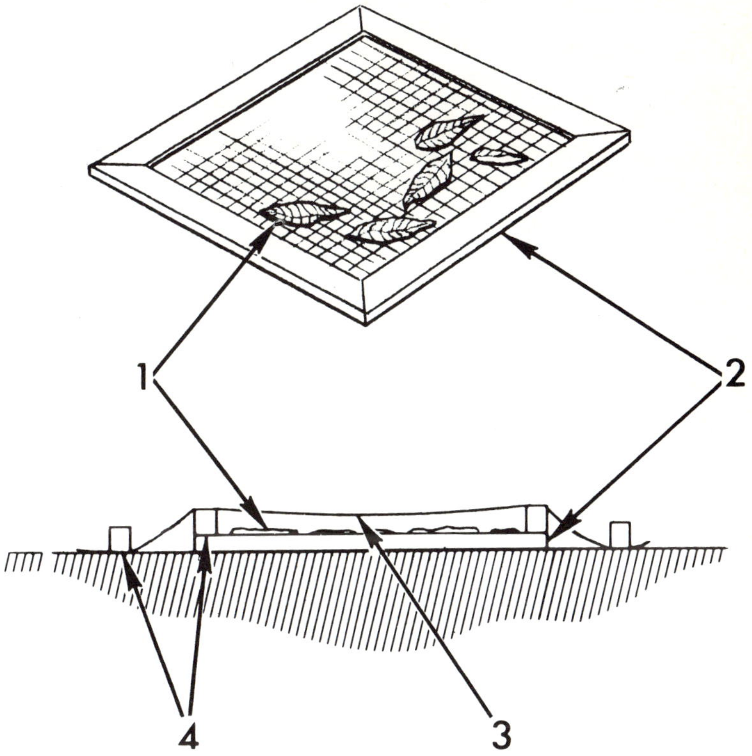

Drying Herbs

Another way of drying herbs is to hang the herbs in small bunches [1], leaves down. With the leaves down, the oils from the stems flow into the leaves.

To protect the leaves from dust and from direct light, tie a paper bag [2] around the leaves. Punch holes [3] in the bag to allow air to circulate.

For dill, the seeds will drop into the paper bag as they dry.

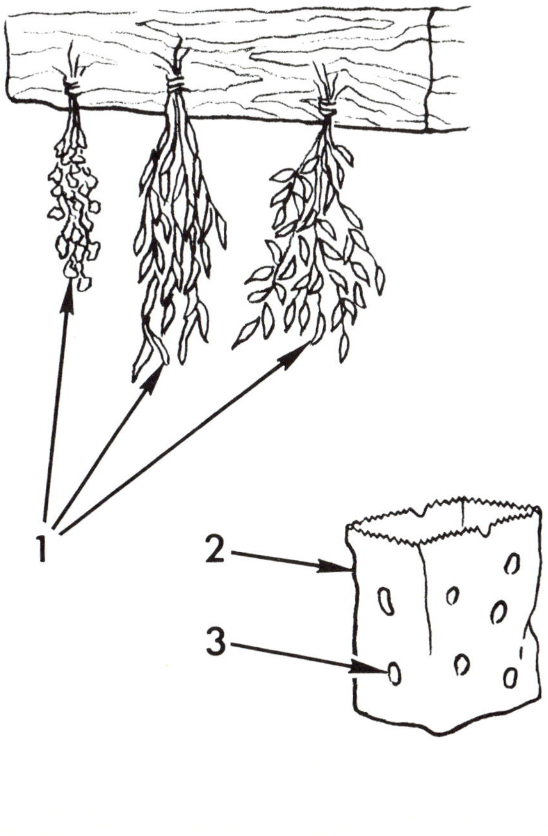

▶ Storing Herbs

After drying, store the herbs in tightly-sealed jars for about a week. At the end of this time, check for moisture on the inside of the jar. If there is moisture, remove the herbs and spread them out for further drying.

Do not crush leaves or grind seeds before storing. They retain their oils and flavor much better if kept whole. Crush them just before use.

Do not store herbs in paper or cardboard containers, as the oils will be absorbed and the herbs lose their flavor.

Store the herbs in glass jars, in a cool place away from strong light.

STORING VEGETABLES

Most vegetables can be stored for a period of time without freezing, canning, or drying. Some are much easier to store than others and can be stored for a longer time.

Vegetables can be stored in boxes [1], baskets [2], crates [4], etc. The location can be a basement or cellar [3], outbuilding, underground pit, and other possibilities.

The main requirement for most vegetables is a cold, moist environment. The table on the next page shows the approximate temperature and humidity which is best for storing each vegetable. Note that except for onions and winter squash, high humidity is best for storage. The approximate length of time for storage for each vegetable is also shown.

While the best storage temperature shown for many vegetables is 32°F, it is better to aim for a temperature just above this (35–40°F) to avoid freezing the vegetable. Vegetables (except potatoes) are not damaged at 32°F, but it is difficult to maintain this low a temperature without actually freezing the vegetables.

Unfortunately, we can give no hard and fast rules to achieve these storage conditions. Climatic and other conditions vary too greatly. Some guidelines, however, are given in the following pages.

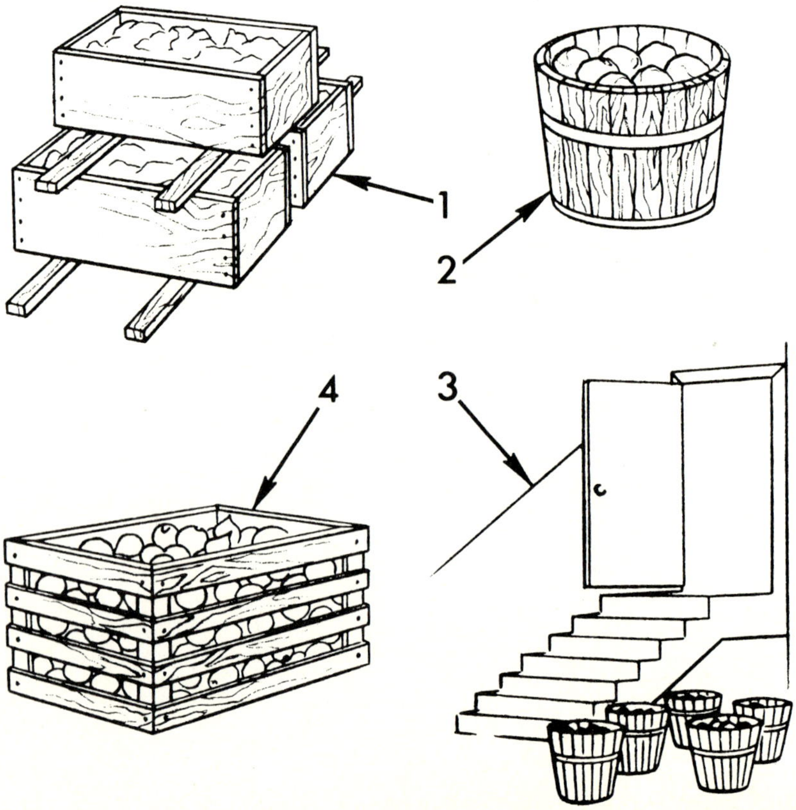

BEST STORAGE CONDITIONS FOR VEGETABLES			
Vegetable	**Temperature (°F)**	**Relative Humidity**	**Storage Life (approx.)**
Asparagus	32–36	95%	2–3 weeks
Beans			
Snap	40–50	95%	7–10 days
Lima	40	90%	1–2 weeks
Beets	32	95%	3–5 months
Broccoli	32	95%	10–14 days
Brussels sprouts	32	95%	3–5 weeks
Cabbage	32	95%	1–3 months
Carrots	32	95%	4–5 months
Cauliflower	32	95%	2–4 weeks
Celery	32	95%	2–3 months
Corn	32	95%	4–8 days
Cucumbers	45–50	95%	10–14 days
Eggplant	45–50	90%	1 week
Lettuce	32	95%	2–3 weeks
Melons			
Cantaloupe	40	90%	2 weeks
Watermelon	40–50	80–85%	2–3 weeks
Onions, bulb	32	60–70%	6–7 months
Peas	32	95%	1–3 weeks
Peppers, sweet	45–50	95%	2–3 weeks
Potatoes	40	90%	4–9 weeks
Radishes	32	95%	3–4 weeks
Spinach	32	95%	10–14 days
Squash, winter	50–55	50–60%	2–6 months
Tomatoes, green	50–70	90%	1–3 weeks
Tomatoes, ripe	45–50	90%	4–7 days
Turnips	32	95%	2–4 months

▶ Storing Varieties

As for freezing and canning, certain varieties are especially good for storing. These varieties are shown in the accompanying table.

The companies from which the varieties can be ordered are shown on the right hand side of the table. Wherever an "X" appears, you can order that variety from the company above the X. Addresses of the companies are given on Page 61.

GOOD VEGETABLE VARIETIES FOR STORING

Vegetable	Varieties	Burgess	Burpee	Farmer	Field	Gurney	Harris	Jung	May	Shumway
Beets	Detroit Dark Red	X	X	X	X	X	X	X	X	X
Cabbage	Autumn King									X
	Copenhagen Market	X	X		X				X	X
	Danish Ballhead	X	X		X	X	X	X	X	
	Early Flat Dutch		X	X	X					X
	Jumbo	X								
	Mammoth Red Rock			X		X			X	
	Penn State Ballhead			X						
	Premium Late Flat Dutch			X	X					X
	Surehead		X		X					
	Wisconsin All Season			X						X
	Wisconsin Red Hollander			X						X
Carrots	Danvers Half Long		X	X	X	X	X			X
	Gold Pak			X	X	X		X		X
Celery	Fordhook		X							
	Giant Pascal		X	X						X
	Golden Self-Blanching		X		X	X			X	X
Onions bulb	Downing Yellow Globe					X	X			
	Ebenezer	X	X			X	X			X
	Hybrid Yellow Sweet Squash	X	X	X	X	X	X		X	X
	Red Wethersfield Hamburger									X
	Ringmaster	X				X				
	Southport Red, White and Yellow Globes			X		X	X	X	X	X
	White Sweet Keeper					X				
	White Sweet Slicer					X				
	White Sweet Spanish					X				X

Storing Varieties

GOOD VEGETABLE VARIETIES FOR STORING (contd)

Vegetable	Varieties	Burgess	Burpee	Farmer	Field	Gurney	Harris	Jung	May	Shumway
Potatoes	Kennebec			X	X	X			X	
	Norgold Russet			X	X	X				
	Red Pontiac			X	X	X			X	
	White Cobbler				X	X			X	
Radishes	Cherry Belle			X	X		X	X	X	X
	China Rose				X			X	X	X
	Comet		X							
	Long Black Spanish									X
	Round Black Spanish			X		X				X
	Round White	X								
	Stop Lite			X						
	White Chinese	X	X							
	White Giant				X			X		X
Squash Winter	Buttercup	X	X	X		X	X	X	X	X
	Butternut	X	X	X		X	X	X	X	
	Gold Nugget			X		X	X	X	X	
	Golden Delicious					X	X			
	Hubbard (Golden or Red)	X	X	X						
	Hubbard (True or Warted)	X	X	X	X		X	X	X	
	Royal Acorn		X							
	Table Queen	X	X	X	X	X	X	X	X	X
	Waltham Butternut		X	X	X			X		
Turnips	Purple Top White Globe	X	X	X	X	X	X	X	X	

▶ Preparing Vegetables

Select only good quality vegetables for storage. Vegetables which are damaged can be eaten immediately, or the bad parts can be cut out and the vegetables frozen or canned.

Vegetables should be washed first, but let the water dry before storing.

If you desire, you can place vegetables in plastic bags [1] before packing in a container. The bags should have holes for ventilation. The bags will reduce the rate of wilting by retaining moisture.

Root crops—beets, carrots, and turnips—can be left in the ground where they grew. Later, dig up with a spading fork [3]. For ease of digging up, cover the vegetables with a mulch [4] of hay or leaves.

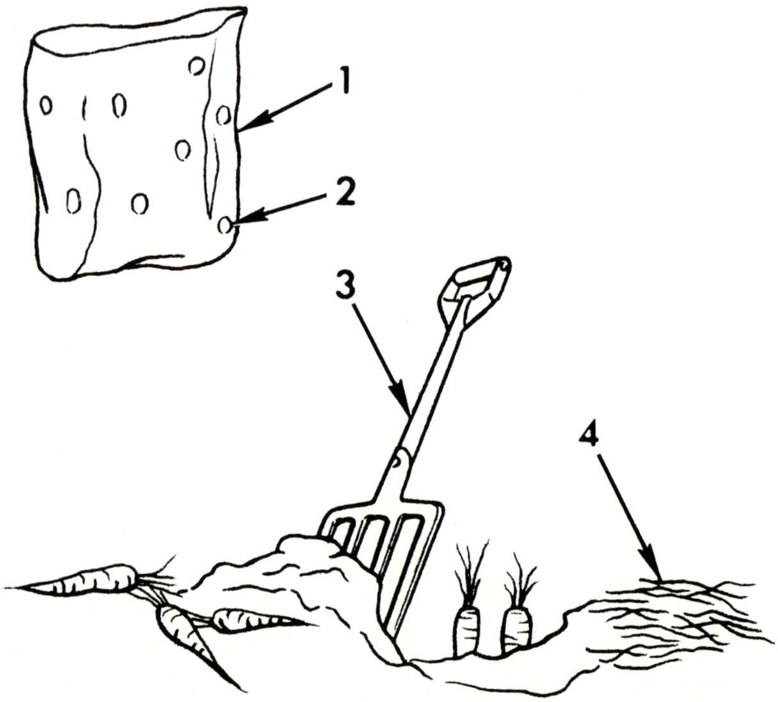

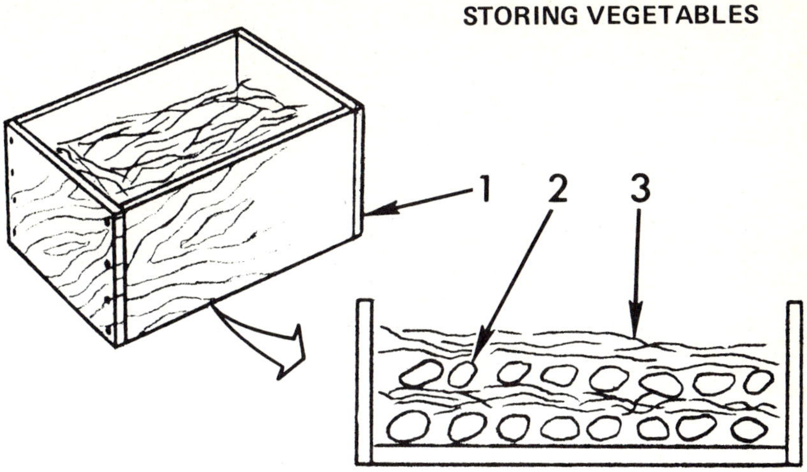

▶ Packing in Containers

Wooden boxes [1] are good storage containers. Use an insulating material [3] such as dry leaves, burlap, hay, straw, or sphagnum moss. Lay alternate layers of vegetable [2] and insulating material [3] in the box.

Barrels, pails, or baskets can be used in the same manner as boxes. Fill with alternate layers of vegetables and insulating material.

Onions [4] do not need insulating material. They are best stored in crates [5], which provide good air circulation.

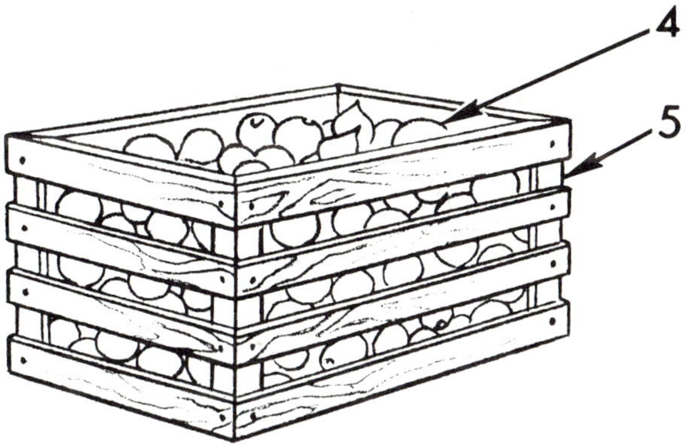

▶ Storage Location

The storage area should be kept clean at all times. Do not allow dead leaves, stalks, etc., from vegetables to accumulate. Keep ceiling, walls, and floor clean and free of spiders and other insects.

Most vegetables shrivel rapidly unless stored in a moist environment.

One good way or providing moisture in the storage room is as follows:

1. Spread a 2-inch layer of gravel [2] on the floor.

2. Place boxes [1] and other containers on the floor. Use wooden runners [3] to separate boxes and to keep boxes off the floor.

Avoid causing puddles when watering.

3. Using a watering can, sprinkle water on the gravel.

The water will evaporate from the gravel, making the air moist. Keep the watering can handy, for periodic watering.

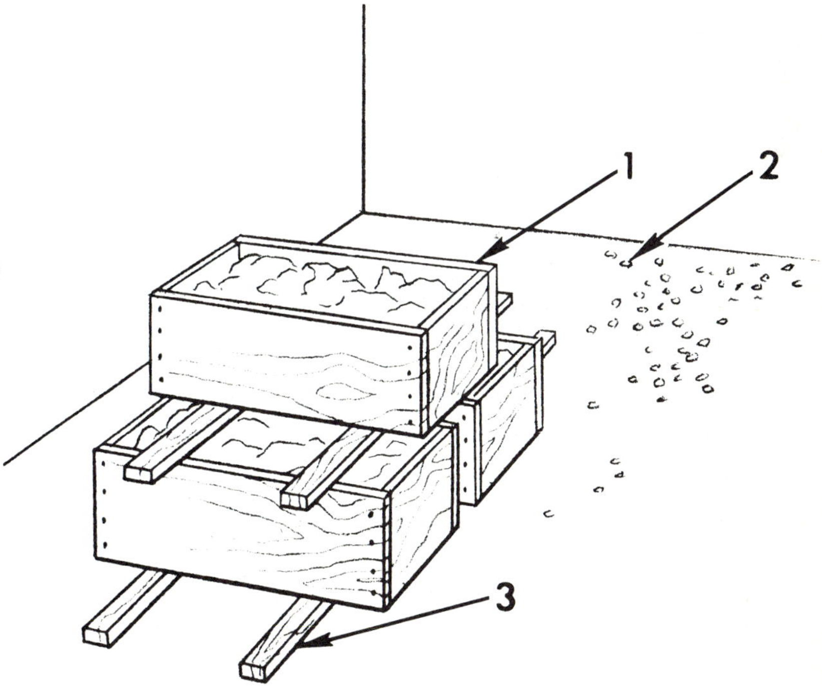

STORING VEGETABLES

Storage Location

Instead of using a gravel floor for moisture, water may be added directly to vegetables. The watering can is used to sprinkle water as needed. This method is more practical if the vegetables are packed without layering. They are placed in containers and covered by burlap, rugs, etc., which are also moistened periodically.

Basements or cellars are good places for storing vegetables. There is usually an outside entrance which can be used for ventilation. There is often an insulation material on the ceiling to prevent cold air in the cellar from cooling the house.

Basements with cement floors may often be too warm for vegetable storage, without modification. This is particularly the case if there is central heating. However, a corner of the basement can be converted to a storage area by erecting insulating walls, such as concrete block. A northeast or northwest corner of the basement is usually best for this purpose. There should be a window for letting cold air in when needed, to keep the temperature down around 32°F.

Many gardeners have found that the ground is a good place for storing vegetables. However, if pit-type storage [1] is used, be sure to use wire screen [2], hardware cloth, or similar material for protection from rodents.

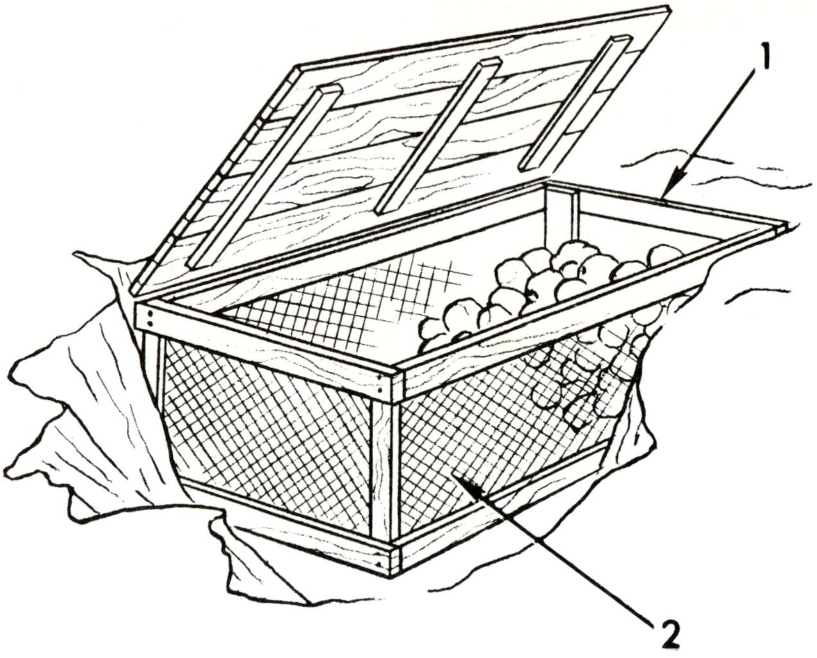

▶ Checking Vegetables

Stored vegetables should be checked from time to time. Look for shriveling, decay, or growth.

Remove and discard any decayed vegetables.

Vegetables which start to shrivel should be placed in vented plastic bags, or moisture should be applied by direct sprinkling or by sprinkling the ground floor (see Page 85).

If vegetables show growth, the temperature is too high. Check the table on Page 83 for the best temperatures to store different vegetables.

▶ Special Requirements of Certain Vegetables

Certain vegetables have special storage requirements which should be considered.

If both fruit and vegetables are being stored, don't store apples and pears with cabbage, potatoes, or turnips. Apples produce a gas which will cause potatoes to sprout. Cabbage and turnips produce odors which become attached to apples and pears.

Beets, carrots, turnips (root crops): may be left in the ground for storage where they grew. To make it easier to dig them up, cover with a mulch of hay, straw, or leaves.

Cabbage: store in boxes or bins in cold, damp area, or underground in damp sand or soil. If to be stored in boxes or bins, remove roots and stem and wrap with newspaper, burlap, or similar material. If to be stored underground, leave roots and stem attached.

Celery: in areas without severe winters, celery may be left in the ground and covered with a heavy mulch of hay, straw, or leaves.

Onions (bulb): onions must be "cured" by leaving on the ground for a few days after pulling, then placing in crates in a ventilated area for several weeks. The tops are then removed and the onions are stored in boxes or bins under the conditions shown in table on Page 83. The right storage conditions for onions are often found in attics.

Potatoes: a dark area is needed for storage. Potatoes may be held for several months after harvest under a wide range of storage conditions. To prevent sprouting, however, the potatoes should eventually be stored under conditions shown in table on Page 83.

Squash (winter): like onions, need to be "cured". Leave short stem on when harvesting, and let squash remain on ground for 2 weeks. Then store them under the conditions shown in table on Page 83.

NOTES

NOTES

NOTES

NOTES

NOTES

NOTES

NOTES

NOTES

NOTES

NOTES

NOTES

NOTES

NOTES

NOTES

NOTES

NOTES